Engineered Repairs of Composite Structures

Engineered Repairs of Composite Structures

Rikard Benton Heslehurst

CRC Press
Taylor & Francis Group
Boca Raton London New York

CRC Press is an imprint of the
Taylor & Francis Group, an **informa** business

CRC Press
Taylor & Francis Group
6000 Broken Sound Parkway NW, Suite 300
Boca Raton, FL 33487-2742

First issued in paperback 2020

ISBN-13: 978-1-4987-2626-9 (hbk)
ISBN-13: 978-0-367-77996-2 (pbk)

Library of Congress Cataloging-in-Publication Data

Names: Heslehurst, Rikard Benton, author.
Title: Engineered repairs of composite structures / Rikard Benton Heslehurst.
Description: First edition. | Boca Raton, FL : CRC Press/Taylor & Francis
Group, 2018. | Includes bibliographical references and index.
Identifiers: LCCN 2018052810 | ISBN 9781498726269 (hardback : acid-free paper)
| ISBN 9780429198656 (ebook)
Subjects: LCSH: Concrete construction--Maintenance and repair. | Buildings,
Reinforced concrete--Maintenance and repair.
Classification: LCC TA682.446 .H47 2018 | DDC 624.1/834--dc23
LC record available at https://lccn.loc.gov/2018052810

Visit the Taylor & Francis Web site at
http://www.taylorandfrancis.com

and the CRC Press Web site at
http://www.crcpress.com

I dedicate this book to the important women in my life: my wife Fiona, my daughter Elyse, my mum Jan, my sister Leanne, my mother-in-law Jean, and both of my grandmothers Nan (Edna) and Mumma (Phyllis). They all are responsible for the person I have become, and I am grateful for their influence.

Contents

Preface

The aim of this book on engineered repair to composite structures is to provide a detailed discussion, analysis and procedures for effective and efficient repair design. The book discusses the identification of damage types and the effect on structural integrity in composite structures. This leads to the design of a repair scheme that suits the restoration of the structural integrity and damage tolerance of the composite structure. The book will also discuss the application of the repair and what is needed in the forming of the composite repair to meet the engineering design requirements.

Composite materials have many advantages over conventional structural materials, particularly in primary structures. The advantages of composite materials include a high specific strength and stiffness, formability and a comparative resistance to fatigue cracking and corrosion. However, not forsaking these advantages, composite materials are prone to a wide range of defects and damage during manufacture and in-service. The damage can significantly reduce the residual strength and stiffness of the structure or result in unfavourable load paths.

When the damage is found, the next step is to determine the loss of structural or performance integrity. The stress analysis of damaged areas considers the four generic damage types: intralaminar matrix cracks, matrix degradation due to burning and chemical attack, interlaminar matrix cracks (delaminations) and fibre fracture (holes). The damage criticality then allows a better approach to the repair scheme. Like the damage stress analysis, the repair scheme is often driven by the damage type. Each damage type can have a unique repair scheme, but the design of the repair scheme needs to account for the impact of material stiffness changes. Since the repair is a structural joint, the load transfer from parent structure to repair material must be understood in terms of joint stresses and joint geometry.

The physicality of removing damage from composite structures, preparing the damage surface for repair, fabricating the repair scheme and installing the repair scheme is discussed in detail. Several guidelines and suggestions are provided that will greatly assist in the physical installation of a repair that better meets the engineering requirements of the damaged composite structure restoration.

The application of the intent of this book will allow for more structurally efficient and effective repairs of damaged composite structures.

Acknowledgments

The work described in this book could not have been possible except with the assistance from the following people and organisations:

- Mike Hoke, Lou Dorworth, Dave Castellar, Mike Meringolo and Jay Carpenter.
- The staff at University College (UNSW), at the Australian Defence Force Academy.
- My friends at SAMPE and Composites Australia.

Author

Rikard Benton Heslehurst, PhD (UNSW) is a former aeronautical engineering officer of the Royal Australian Air Force (RAAF). He is now the CEO and director of his own RDT&E company, Heslehurst and Associates Pty., Ltd. in Garran, Australia and Chief Engineer for M51 Resources Inc, Dallas, Texas. Rik recently retired as a senior lecturer with the University of New South Wales at the Australian Defence Force Academy, Canberra (now an honorary academic) and was formerly the senior engineer at Abaris Training, Reno, Nevada. Rik is a Chartered Professional Engineer, a Fellow of the Institute of Engineers, Australia, a Fellow of the Royal Aeronautical Society and a SAMPE Fellow.

Nomenclature

ACRONYMS

AFFDL	Air Force Flight Dynamics Laboratory (US)
AFRP	Aramid Fibre Reinforced Plastic
AFWAL	Air Force Wright Aeronautical Laboratories (US)
AGARD	Advisory Group for Aerospace Research & Development (NATO)
AGPS	Australian Government Publications
AIAA	American Institute of Aeronautics and Astronautics
ASM	American Society of Materials
ASTM	American Society for Testing and Materials
BFRP	Boron Fibre Reinforced Plastic
BVID	Barely Visible Impact Damage
CDS	Characteristic Damage State
CFRP	Carbon (Graphite) Fibre Reinforced Plastic
CMH	Composite Materials Handbook
CTE	Coefficient of Thermal Expansion
CTHERM	Coefficient of Thermal Expansion Parameter
DOT	Department of Transport (US)
DST	Defence Science Technology Group (AUS)
DSTO	Defence Science Technology Organisation (now DST) (AUS)
ESA	European Space Agency
ETR	Stiffness–Thickness Ratio
FAA	Federal Aviation Administrative (US)
FEA	Finite Element Analysis
FoS	Factor of Safety
FPF	First Ply Failure
GFRP	Glass Fibre Reinforced Plastic
IAW	In Accordance With
LARC	Langley Research Center (NASA)
LEFM	Linear Elastic Fracture Mechanics
MIL-HDBK	Military Handbook
MMPDS	Metallic Materials Properties Development and Standardisation
NADC	Naval Air Development Center (US)
NAEC	Naval Air Engineering Center (US)
NASA	National Aeronautics and Space Administration (US)
NDE	Non-Destructive Evaluation
NDI	Non-Destructive Inspection
OEM	Original Equipment Manufacturer
QI	Quasi-Isotropic
RAAF	Royal Australian Air Force
RAF	Royal Air Force (UK)
RAE	Royal Aircraft Establishment (UK)

RCS	Radar Cross-Section
RT	Room Temperature
RTO	Research and Technology Organization (NATO)
SAE	Society of Automotive Engineers (US)
SAMPE	*Society for the Advancement of Material and Process Engineering*
SIFT	Strain Invariant Failure Theory
SRM	Structural Repair Manual
STP	Special Technical Publication
T-T-T	Through-the-Thickness

SYMBOLOGY-ALPHA-NUMERIC

$(Et)_{patch}$	effective stiffness of the repair patch
$[A]$	laminate in-plane stiffness matrix (absolute)
$[a]$	laminate in-plane compliance matrix
$[a_{ij}]$	laminate compliance matrix
$[A_{ij}]$	laminate in-plane stiffness matrix
$[B]$	laminate coupling stiffness matrix (absolute)
$[D^*]$	flexural stiffness matrix
$[D]$	laminate flexural stiffness matrix (absolute)
A	total laminate panel surface area ($a \times b$)
a	hole radius, ellipse major axis, diameter of the delamination being assessed, panel length in the 1-direction
A^*	effective area of delaminated region
a'_{ij}	inclusion compliance matrix
A_{ij}	in-plane laminate stiffness matrix
a_{ij}	composite compliance plate
a_o	laminate characteristic dimension
A_{Shear}	fastener shear force area
b	panel width in the 2-direction, width of a beam or panel, ellipse minor axis
b_o	characteristic dimension
C	constant or proportionality
C_f	fastener effective compliance
CTHERM	thermal expansion parameter
D	fastener diameter
D^*	reduced flexural stiffness matrix
D_1	flexural stiffness for isotropic materials
D_{11}	flexural stiffness for orthotropic materials
E_f/E_m	Degree of Anisotropy
D_{hole}	hole diameter
D_{ij}	flexural stiffness matrix components
d_o	characteristics dimension
DofO	degree of orthotropy
D_w	washer diameter
E	stiffness parameter

e	bolt edge distance, eccentric distance
E_{f1}	composite longitudinal flexural Young's modulus
E^*	effective stiffness of the multiple delaminations
E_c'	effective peeling Young's modulus
E_1	longitudinal Young's modulus of the laminate
E_2	transverse Young's modulus of the laminate
E_c	peeling Young's modulus
E_{damage}	damaged region stiffness
E_f	fastener modulus of elasticity
E_i	effective Young's modulus of the ith sub-laminate
E_{lam}	laminate axial Young's modulus
E_{o_1}	in-plane laminate longitudinal Young's modulus
$E_o t_o$	outer adherend effective stiffness
E_{parent}	parent laminate effective principle stiffness
E_{patch}	repair patch stiffness
$E_{\text{SL}_{ni}}$	effective sub-laminate in-plane stiffnesses
E_x	longitudinal Young's modulus
E_y	transverse Young's modulus
E_z	through-the-thickness Young's modulus
F_{ij}	strength criteria parameter
f_j	orthotropic material stress concentration parameter
FoS	factor of safety
F_u	ultimate adherend strength
G	shear stiffness in the linear region, strain energy release rate
G_{12}	composite in-plane shear modulus
G_{I}	mode I strain energy release rate
G_{Ic}	critical mode I strain energy release rate
G_{II}	mode II strain energy release rate
G_{III}	mode III strain energy release rate
G_{init}	initial shear modulus
g_j	orthotropic material stress concentration parameter
G_{xy}	in-plane shear modulus
G_{xz}	interlaminar shear modulus
G_{yz}	interlaminar shear modulus
h	laminate thickness
h_a	adhesive thickness
k	single lap joint eccentricity parameter, biaxial loading parameter
K	plate stiffness (AE/l)
K_f	fastener spring stiffness
K_T	orthotropic material stress concentration parameter
k_{tc}	composite net-tension stress concentration factor
k_{te}	elastic isotropic net-tension stress concentration factor
l	length or component or joint overlap length
l_{overlap}	adhesive joint overlap length
l_p	patch length
l_{patch}	patch length

m	number of delaminations at a particular through-the-thickness point, number of strips or fasteners in a row, number of sub-laminates, rate of degradation, integer values representing the half sinewave number for critical buckling mode.
n	sub-laminate number, number of shear planes, integer values representing the half sinewave number for critical buckling mode
N_i	applied in-plane axial and shear loads
N_1^{cr}	critical applied in-plane axial
p	Pitch Distance
$\mathbf{P}$	Load in joint
$P_{\text{adhesive-shear}}$	shear load in adhesive
P_{all}	allowable load per unit width of the patch
$P_{\text{Allowable}}$	allowable axial force
P_{fastener}	fastener force
R	hole radius, panel aspect ratio
R_{FPF}	first ply failure strength ratio
R_{Ult}	ultimate ply failure strength ratio
S	composite in-plane shear strength, ply ultimate shear strength, strain energy density
S_C	critical strain energy density
S_{min}	minimum strain energy density
$\mathbf{T}$	torque (including washer), maximum life, interlaminar shear strength
t	thickness of structure, laminate thickness
t_{cs}	equivalent cold storage time
t_{damage}	depth of damaged region
t_i	thickness of the ith sub-laminate
t_o	out the freezer for time
t_{patch}	patch thickness
t_{repair}	thickness of repaired region
V_f	fibre volume ratio
w	strip width, width of specimen or fastener seam spacing
X	x-axis composite ply ultimate strength
x	longitudinal axis direction
X_C	composite longitudinal compression strength
X_T	composite longitudinal tensile strength
xy	in-plane
xz	interlaminar plane
Y	y-axis composite ply ultimate strength
y	transverse axis direction
Y_C	composite transverse compression strength
Y_T	composite transverse tensile strength
yz	interlaminar plane
Z	z-axis ultimate composite strength
z	normal axis direction

GREEK

h_a	adhesive thickness
Γ	circumferential hole stress parameter
Ω	angle of an elliptical hole axis from the 2-axis direction
α	hole stress concentration parameter, elliptical hole aspect ratio, fastener width-to-diameter ratio, CTE
β	hole stress concentration parameter, fastener edge distance-to-width distance ratio
χ	bondline length parameter
δ	hole stress concentration parameter
ε_{all}	allowable design strain
ε_c	peel strain
ε_{limit}	limit design strain
ε_x	axial strain in the x-direction
ε_y	axial strain in the y-direction
ε_z	axial strains in the z-direction, normal strain
γ_e	the elastic shear strain
γ_{max}	maximum shear strain
γ_p	the plastic shear strain
γ_{xy}	shear strain in the xy-plane, in-plane shear strain
γ_{xz}	shear strain in the yz-plane, interlaminar shear strain
γ_{yz}	shear strain in the yz-plane, interlaminar shear strain
κ	circumferential hole compliance parameter
λ	hole stress concentration parameter, elliptical hole aspect ratio, overlap length parameter
μ	hole stress concentration parameter, laminate characteristic equation complex roots
V	circumferential hole compliance parameter
θ	angle from the x-axis, positive CCW, fastener edge distance parameter, scarf angle, ply orientation
θ_{min}	minimum scarf angle
ρ	hole radius off-set parameter, hole stress concentration parameter
σ_1	axial stress in the 1-direction
σ_2	axial stress in the 2-direction
$\sigma_{adhesive}$	through-the-thickness stress in the adhesive
σ_{axial}	axial stress
$\sigma_{bending_{max}}$	bending stress
σ_{BR}	bearing stress/strength
σ_c	peel strength
$\sigma_{c_{max}}$	maximum peel stress
σ_i	far-field in-plane stresses for $i = 1,2,6$
σ_{max}	maximum axial stress
σ_N	axial notch stress
σ_{NT}	net tension stress/strength
σ_{o1}	axial stress in the primary structural direction (1)

σ_x	axial stress in the x-direction
σ_y	axial stress in the y-direction
σ_z	axial stress in the z-direction, through-the-thickness normal stress
σ_θ	circumferential stress around a hole boundary
τ_{12}	shear stress in the 1-2 plane
$\tau_{adhesive}$	shear stress in the adhesive
τ_p	plastic shear stress (maximum shear stress)
τ_{shear}	shear stress in a fastener
τ_{SO}	shear-out stress/strength
τ_{xy}	shear stress in the xy-plane
τ_{xz}	shear stress in the xz-plane, interlaminar stress components
τ_{yz}	shear stress in the yz-plane, interlaminar stress components
υ_{12}	laminate minor Poisson's ratio
υ_{21}	laminate major Poisson's ratio
ξ	bondline geometry parameter
ψ	angle around the ellipse CCW from the +ve x-axis

SUPERSCRIPTS AND SUBSCRIPTS

*	modified value
'	effective value, variational value
1	1-axis direction (longitudinal)
12	in-plane value
2	2-axis direction (transverse)
21	value change in 2-direction due to effect of value in 1-direction
3	3-axis direction (through-the-thickness)
a	adhesive
all	allowable value
bc	composite bearing
br	bearing
C	compression
c	compression, critical value, peel value
cr	critical value
cs	cold storage value
damage	damaged region
e	elastic
f	fastener, fibre, flexural value
FPF	*first ply failure value*
hole	hole in a structure
i	inner adherend (or for repair design equivalent to p ... parent structure), in-plane value, numerical index
I	Mode I opening behaviour
II	Mode II opening behaviour
III	Mode III opening behaviour
init	initial value
j	numerical index

limit	limit value
m	matrix (resin), maximum count number, index power
max	maximum value
min	minimum value
N	maximum notch value
n	buckling value, index power, local notch value
NT(nt)	net tension
o	outer adherend (or for repair design *r* … repair patch), original or origin value, out value
p	index power, plastic, repair patch
parent	parent value
patch	repair patch
repair	repair region
so	shear-out
T	tension, total value
t	tension
tc	composite net tension
te	elastic isotropic net tension
u	ultimate value
Ult	ultimate value
w	washer
x	coordinate longitudinal axis
xy	coordinate axes in-plane shear face
xz	coordinate axes normal shear face
y	coordinate transverse axis
yz	coordinate axes normal shear face
z	coordinate normal axis
θ	angular direction

1 Introduction

Composite materials have many advantages over conventional structural materials, particularly in primary structure. The advantages of composite materials include a high specific strength and stiffness, formability, and a comparative resistance to fatigue cracking and corrosion. However, not forsaking these advantages, composite materials are prone to a wide range of defects and damage during manufacture and in-service. The damage can significantly reduce the residual strength and stiffness of the structure or result in unfavourable load paths. Damage and defects can also reduce the functional performance of the composite component, such as leaking of pressure vessels and liquid tanks.

When the damage is found in the composite structure (termed damage assessment), the next step is to determine the loss of structural or performance integrity (termed damage analysis). Details of the stress analysis of damaged areas will be different for the damage type found. The loss of structural integrity and the methods used to determine this loss of structural integrity will be based on whether the damage is intralaminar matrix cracks, matrix (resin) burnout, interlaminar matrix cracks (delaminations), fibre fracture (holes) or a combination of any of these. The determination of the damage criticality will then allow for a better approach to the design and development of the repair scheme. Like the selected damage stress-analysis method, the repair scheme is often driven by the damage type. Each damage type can have a unique repair scheme, but the design of the repair scheme needs to account for the impact of stiffness changes in the parent structure. Since the repair is a form of joint, the load transfer from parent structure to repair material must be understood well in terms of parent structure and joint stresses, and overall joint geometry.

This book of the engineering design of composite structural repairs aims to describe the process of damaged composite structure restoration through identification of the damage type, damage criticality analysis, repair scheme design and repair scheme installation. The numerical assessment of the damage criticality and repair sizing is detailed and supported with examples. The application of the intent of this book will allow for more structure-efficient and -effective repairs of damaged composite structures.

DEFINITION OF COMPOSITE STRUCTURE

A composite material in the context of this book is a material consisting of any combination of filaments and/or particulates in a common matrix. Various material combinations can therefore be called composites, i.e. fibre-reinforced plastics (Figure 1.1).

The basic premise of the term composite materials is that the combination of different materials to form a new material is done in such a way that ensures that each constituent material does not lose its individual form or material properties. The

1

FIGURE 1.1   A fibre-reinforced composite sandwich panel.

composite material is thus a combination of each constituent material and maintains each constituent's own characteristics (both good and poor properties) but provides support to the other constituent to overcome the limitations in the other constituent materials.

In this book, we are predominantly interested in the fibre-reinforced types of composite materials, such that the combined properties of the fibre and matrix are used to enhance one another. The filament or fibre or fabric (Figure 1.2) provides

FIGURE 1.2   The fibres.

the essential axial high strength and stiffness of the composite material, with a low density that gives the significant benefits of exceptionally high specific properties. However, the filament or fibre or fabric is brittle and requires support against premature fracture. Thus, added to the fibre is a matrix (polymer resin, ceramic or metal) (Figure 1.3) that gives good shear behaviour with an ease of fabrication and with a relatively low density. The material is generally more susceptible to defects and damage from the operational environment. Thus, a composite material (structure) is formed by the combination of the fibre and matrix constituents (Figure 1.4). The resulting structure provides an increase in the damage tolerance and toughness of the brittle fibres, with minimal loss of the beneficial mechanical and physical properties of the fibres.

With respect to current composite industry applications, we will confine our discussions to composite materials such as Graphite (Carbon)/Resin (CFRP) and Glass/Resin (GFRP), and polymer/polymer composites such as Aramid/Resin

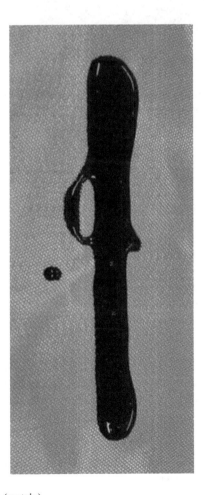

FIGURE 1.3   The resin (matrix).

FIGURE 1.4 The composite material.

(AFRP). The use of Boron/Resin (BFRP) composites is typically reserved for a special range of higher strength and stiffness applications, particularly for compression, but BFRP composites have an important application in repair of metal structures.

TYPICAL COMPOSITE STRUCTURES

Apart from the lamination of the composite materials to define the structure, composites are basically in two forms:

- **Monolithic structures**. Where monolithic structures can be thin-skinned with thin stiffeners (Figure 1.5) or thick structures (Figure 1.6).
- **Sandwich structures**. With relatively thin skins, a light-weight thick structure is formed with a low-density core material (Figure 1.7). Sandwich structures can also have multiple layers of composite and core materials (Figure 1.8).

Another composite monolithic or sandwich structure can use different fibre forms and/or fibre types in the laminate construction (Figure 1.9). Likewise, variations in the core type and/or density can be found in sandwich structures. These structures are commonly termed hybrid composites. A common hybrid is the use of alternative layers of aluminium and glass-fibre/epoxy – called GLARE (Figure 1.10).

All types of the composite structure described above can be damaged and thus will require some form of repair.

FIGURE 1.5 Thin-skinned monolithic composite structure.

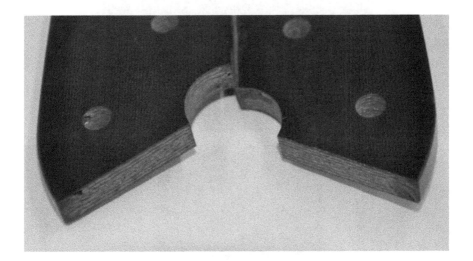

FIGURE 1.6 Thick monolithic composite structure.

FIGURE 1.7 Thin-skinned composite sandwich structure.

FIGURE 1.8 Multi-layered composite sandwich structure.

FIGURE 1.9 Hybrid composite structure.

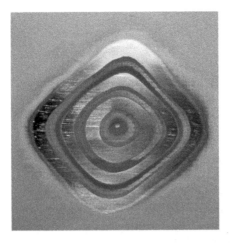

FIGURE 1.10   GLARE composite structure.

DESIGN REQUIREMENTS FOR COMPOSITE STRUCTURES WITH A FOCUS ON THE DAMAGING ENVIRONMENT

The basic design requirements for the development of composite components and structure in a damaging environment can be considered as follows:

- Performance:
  - Functional Performance:
    - Tensile, compressive, shear and/or bearing strength of the composite components and structures
    - Strength loss of composite components and structures due to defects and damage
    - Out-of-plane strength of composite components and structures with and without defects
    - Young's modulus in the orthogonal directions of composite components and structures
    - Axial Young's modulus of fasteners and their interaction with the composite materials
    - Stiffness loss of composite components and structures due to defects and damage
    - Coefficient of thermal expansion of composite components and structures with and without defects
    - Coefficient of moisture absorption of composite components and structures with and without defects
  - Spatial Constraints:
    - Defect or damage size (three-dimensional)
    - Relative defect or damage size to component/structure size
    - Density of defect or damage
    - Location of defect or damage on or in the structure

- Appearance:
  - Effect of defect or damage to the surface profile of the component or structure
- Time:
  - Time to identify defect or damage
  - Time to repair defective or damaged structure
  - Time over which defect or damage has manifested itself
- Cost:
  - Cost of identification of defect or damage
  - Cost of repairing the structure
  - Component or structure downtime cost impact
- Manufacture/Assembly:
  - Repair capability
- Standards:
  - Defect representation
  - NDI personal and equipment
  - Structural repair manual guidance
- Safety of Personnel:
  - During damage inspection
  - During damage repair
- Environmental Issues:
  - Environmental impact due to the existence of the damage or defect
- Maintenance & Repair:
  - Component/structure access
  - Component/structure inspectability
  - Component/structure repairability
  - Facility capabilities
  - Personnel training

OPERATIONAL REQUIREMENTS OF COMPOSITE REPAIRS

The repair of composite structures requires more than just the physical repair action. There are a number of operational requirements that will aid in the successful application of the final repair scheme. These operational requirements are briefly outlined in the following points:

- Identification of the damage (visual inspection [Figure 1.11] system malfunction, operator advise, etc.).
- Assessment of the extent of damage using an approved non-destructive inspection (NDI) procedure, such as a tap test (Figure 1.12). The NDI procedure will provide the size, position and classification of the damage.
- Determination of the damage repair size restriction from the Structural Repair Manual (SRM). If the damage is within the repair size and/or position limits, follow the SRM action. However, if the damage is outside of SRM limits, then seek an engineering disposition on the repair scheme.

- Analysis of the stress state or functional performance degradation of the structure with the damage present. This is done when the damage size and/ or position is outside SRM limits of damage. Stress analysis of the damaged area can be relatively simple or more complex with computation analysis performed to reveal areas of concern, such as finite element analysis (FEA) (Figure 1.13).
- Comparison of the stress state and/or structural performance against structural- or system-design requirements.
  - Obtain the necessary design data from the Original Equipment Manufacturer (OEM) or reverse engineer the structure.
  - Define the damage state repair level or class.
  - Determine the level of repair based on the following:
    - Identification of local repair capabilities, including facilities available, equipment on-site and personnel skill levels.
    - Identification systems integrity control issues.
    - Definition of the repair level/class.
    - Designing of the repair scheme through:
      - the selection of the repair joint,
      - location of the damage in the structure and structural configuration,
      - assessment of the repair installation difficulty due to repair access,
      - determination if a single-sided repair can only be attempted,
      - the ability to undertake numerical analysis of the repair design using appropriate joint design equations,
      - determination of the local facility repair installation capabilities, and
      - assessment for any repair structure geometric constraints, vertical clearances for installation, part curvature, short edge distances, etc.
- Remove the damage (the least amount of undamaged material where possible) based on the following conditions:
  - Geometric constraints of the damaged area.
  - Simple shapes – circular holes or round-ended (domed) rectangles (Figure 1.14).
- Repair area preparation through the following actions:
  - Match drill doubler and parent fastener holes.
  - Metal surface preparation IAW with standard practices.
  - Composite surface preparation IAW with standard practices.
- Repair scheme installation.
- Post-repair action, such as inspection and functionality checks, bolt/surface condition and bolt group alignment (Figure 1.15).
- Paint the repaired surface IAW with repair instructions.
- Reporting the repair action with repair details in the aircraft logbook.
- Review the repair scheme after the first flight to observe any abnormalities.

FIGURE 1.11   Visual inspection.

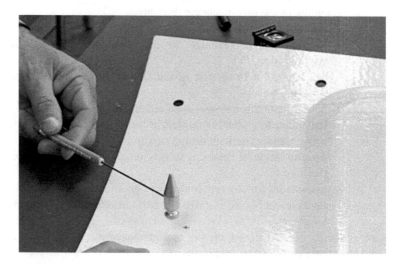

FIGURE 1.12   Assessment of the damaged region.

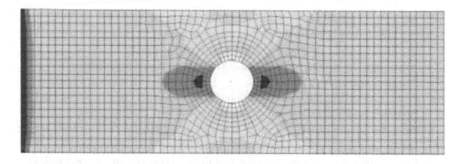

FIGURE 1.13   FEA of a structure with hole damage.

FIGURE 1.14   Round-ended (domed) rectangular cut-out.

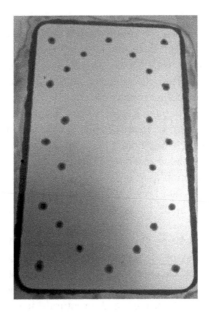

FIGURE 1.15   Bolt group alignment.

WHAT THIS BOOK CONTAINS

This book on the engineered repairs of composite structures is divided into nine chapters. After Chapter 1: Introduction, the following chapters are briefly described:

- Chapter 2: Damage and Defect Types in Composites. Chapter 2 provides a summary of the many damage and defect types found in composite structures, when non-conformity occurs in the lifecycle of the composite structure, the size effect of the non-conformity, the typical location of the non-conformity in the non-conformity and how to generalise a non-conformity in composite structures.
- Chapter 3: Damage Assessment. Damage assessment of composite defects is simply the art and science of finding the non-conformity in composite materials via typical non-destructive inspection (NDI) techniques. A brief review of the principles of NDI is given and then an overview of the

principal NDI methods used in composite structures. Mapping requirements of the non-conformity are defined for the next stage of the repair process, damage analysis.

- Chapter 4: Damage Stress Analysis. For successful damage analysis, i.e. a determination of the loss of structural and/or performance integrity, a viable failure criterion is required. Stress analysis of generalised damage and defects, i.e. intralaminar matrix cracks, heat damage, interlaminar matrix cracks (delaminations), holes and broken fibres are then determined. The damage size and damage density effects are discussed in defining the repair level.

- Chapter 5: Generic Repair Schemes. Chapter 5 gives an overview of fundamental repair schemes, their advantages and disadvantages, and when/ where/how to implement these simple generic repair schemes. The generic repair schemes are considered as: surface damage repair, doubler repair (bonded and bolted), laminate replacement (bonded-scarf repair) and thin-skinned sandwich structure repair.

- Chapter 6: Joining Types and Requirements. All repairs to composite structure require some form of joint. Chapter 6 will review the stress analysis associated with bonded and bolted joints in repair. First, the consideration of which is best, a bonded or a bolted repair is discussed. A thorough overview of several common joint types is given, including single-strap joints, double-strap joints, scarf joints, stepped-lap joints, along with the general aspects of bonded repair joints, bolted joints and the combined bolted/ bonded joint.

- Chapter 7: Repair Scheme Design. The design of the repair scheme needs a good understanding of the repair design requirements that are discussed above in this chapter. Chapter 7 expands on this list of requirements when considering the four repair types: cosmetic repair schemes, damage tolerance restitution repair schemes, and structural integrity restitution repair schemes by either bonded-scarf repairs or bolted-doubler repairs.

- Chapter 8: Repair Schemes Application. To achieve quality repairs the installation of the repair patch is as important as the design of the repair scheme. Chapter 8 considers several aspects of repair scheme installation. First is the removal of the damage, then preparation of the damaged area for the repair and the repair scheme itself, and then finally repair application and attachment.

- Chapter 9: Post-Repair Application Quality Assurance. The final chapter of this book is the quality assessment checks of the completed repair scheme. How to ensure that the repair is installed as designed, initial installed integrity and long-term quality checks are discussed.

2 Damage and Defect Types in Composites

INTRODUCTION

Composites and adhesively bonded structures have many advantages over conventional aircraft materials and construction methods, particularly in primary structures. These advantages include a high specific strength and stiffness, formability and a comparative resistance to fatigue cracking and corrosion. However, not forsaking these advantages, these materials and construction methods are prone to a wide range of defects and damage, which can significantly reduce the residual strength of the components. In this chapter is a review of the types of defects and damage found in composite structures, their failure modes and mechanisms, and a general representation of the defect.

The basic description of a composite structure must be restated since this is very important for the understanding of defects and damage in composites. Composite stricture is a heterogenous material in that the fibre and resin system are independent materials. This means that defects and damage can be considered as either a fibre or a resin problem, or an interfacial issue. Also, composite structures are orthotropic (even anisotropic); in other words, fibre directionality affects the concerns of the damage criticality. The orientation of the defect or damage with respect to the fibre direction will influence the structural performance changes. Finally, a composite structure is a laminated structure. The layered nature of composite materials can result in a structurally significant loss of functional and structural performance.

SUMMARY OF DAMAGE AND DEFECT TYPES

DEFINITIONS

A definition of what constitutes a defect and failure are given in the following paragraphs.

Defect

A defect, also known as a discontinuity, a flaw or as damage, is defined as; 'any unintentional local variation in the physical state or mechanical properties which may affect the structural behaviour of the component'.

Failure

Failure of a component or structure is defined as; 'when a component or structure is unable to perform its *primary* function adequately'.

DEFECT TYPES

There are 52 separate defect types that occur in composite components. These 52 defect types range from microscopic fibre faults to large, gross impact damage. Here, a summary of the defect is provided, and then these defects are categorised into specific groups that will assist in the identification of those defects that are of particular concern to the in-service life of composite structures.

DEFECT LISTING

The 52 individual defects found in composite components and structures are listed in Table 2.1 in alphabetical order.

CLASSIFICATIONS OF DEFECT TYPES

Defects and damage in composite materials can be grouped into specific categories according to when they arise during the total life of the components, the relative

---

TABLE 2.1

Listing of 52 Defect Types Found in Composite Components and Structures

Bearing Surface Damage	Blistering	Contamination
Corner/Edge Splitting	Corner Crack	Corner Radius Delaminations
Cracks	Creep	Crushing
Cuts and Scratches	Damaged Filaments	Delaminations
Dents	Edge Damage	Erosion
Excessive Ply Overlap	Fastener Holes	Fibre Distribution Variance
Fibre Faults	Fibre Kinks	Fibre/Matrix Debonds
Fibre Misalignment	Fracture	Holes and Penetration
Impact Damage	Marcelled Fibres	Matrix Cracking
Matrix Crazing	Miscollination	Mismatched Parts
Missing Plies	Moisture Pick-up	Non-Uniform Agglomeration of Hardener Agents
Overaged Prepreg	Over- or Under-Cured	Pills or Fuzz Balls
Ply Underlap or Gap	Porosity	Prepreg Variability
Reworked Areas	Surface Damage	Surface Oxidation
Surface Swelling	Thermal Stresses	Translaminar Cracks
Unbond or Debond	Variation in Density	Variation in Fibre Volume Ratio
Variation in Thickness	Voids	Warping
Wrong Materials		

---

size of the defects and damage, the location or origin of the defect/damage within the structure and those defects and damage that produce a similar effect to a known stress state within the component.

WHEN NON-CONFORMITIES OCCUR IN COMPOSITE STRUCTURE

Defects and damage occur to composite components and structures during three fundamental life phases for the composite. These three fundamental phases are: constituent materials, prepreg and weave processing, composite component manufacture and fabrication, and in-service use. Table 2.2 lists the defects and damage from Table 2.1 in the three life phases: materials processing, component manufacture and service life usage. A brief description of the three life phases is provided in the following sub-paragraphs.

- **Materials Processing**. Materials-processing defects occur during the production and preparation of the constituent materials, of a prepreg because of improper storage or quality control and batch certification procedures leading to material variations. Weaving errors in dry fabric composite is also part of the first phase of life defects and damage.
- **Component Manufacture and Fabrication**. Component manufacture- and fabrication-induced defects and damage occur during the lay-up and cure process or during the post-cure/joining machining and assembly of the composite components.
- **In-Service Use**. Defects and damage will occur to in-service composite components through mechanical action or contact with hostile environments, such as impact and handling damage, local overloading, local heating, chemical attack, ultraviolet radiation, battle damage, lightning strikes, acoustic vibration, fatigue or inappropriate repair action.

SIZE EFFECT OF THE NON-CONFORMITY

The size of a defect or damage in composite components and structures has significant bearing on the strength or performance criticality of the component. Therefore, in this section, we categorise the defects and damage in composite structures under the effective sizes of microscopic and macroscopic. Microscopic defects and damage will require a level of visual enhancement to be seen by the naked-eye. Macroscopic defects and damage can be easily seen by the eye if open to the surface. These two sizing categories are shown in Table 2.3.

LOCATION OF THE NON-CONFORMITY

Defects and damage in composite structures and components may be present in isolation originating from structural features such as cut-outs and bolted joints, or a random accumulation resulting from interaction amongst themselves. However, defects and damage tend to concentrate at geometric discontinuities, as illustrated in Figure 2.1. Under the following headings of geometric discontinuities, free

TABLE 2.2
Listing of Materials Processing, Manufacturing and In-Service Defects and Damage

Materials Processing	Manufacture and Fabrication	In-Service
Damaged Filaments	Blistering	Bearing Surface Damage
Fibre Distribution Variance	Contamination	Corner/Edge Crack
Fibre Faults	Corner/Edge Splitting	Corner Radius Delamination
Fibre/Matrix Debonds	Cracks	Creep
Fibre Misalignment	Delaminations	Crushing
Marcelled Fibres	Debond	Cuts and Scratches
Miscollination	Excessive Ply Overlap	Delaminations
Overaged Prepreg	Fastener Holes	Dents
Prepreg Variability	• Elongation	Debond
	• Improper Installation	Edge Damage
	• Improper Seating	Erosion
	• Interference Fitted	Fastener Holes
	• Missing Fasteners	• Elongation
	• Over Torqued	• Improper Installation
	• Pull Through	• Improper Seating
	• Resin-Starved Bearing Surface	• Interference Fitted
	• Tilted Countersink	• Missing Fasteners
	Fibre Kinks	• Over Torqued
	Fibre Misalignment	• Pull Through
	Fracture	• Hole Wear
	Holes	• Removal and Reinstallation
	• Drill Burn	• Tilted Countersink
	• Elongation	Fibre Kinks
	• Exit Delamination	Fracture
	• Misdrilled and Filled	Holes and Penetration
	• Porosity	Matrix Cracking
	• Tilted	Matrix Crazing
	Mismatched Parts	Moisture Pick-up
	Missing Plies	Reworked Areas
	Non-Uniformed Agglomeration of Hardener Agents	Surface Damage
	Over- or Under-Cured	Surface Oxidation
	Pills and Fuzz Balls	Surface Swelling
	Ply Overlap or Gap	Translaminar Cracks
	Porosity	
	Surface Damage	
	Thermal Stresses	
	Variation in Density	
	Variation in Resin Fraction	
	Variation in Thickness	
	Voids	
	Warping	
	Wrong Materials	

TABLE 2.3
Listing of Defects and Damage Based on Their Relative Size

Microscopic	Macroscopic
Contamination	Bearing Surface Damage
Creep	Blistering
Damaged Filaments	Contamination
Fibre Distribution Variance	Corner/Edge Splitting
Fibre Faults	Corner Cracks
Fibre/Matrix Debonds	Corner Radius Delamination
Fibre Misalignment	Cracks
Marcelled Fibres	Crushing
Matrix Cracking	Cuts and Scratches
Miscollination	Delaminations
Moisture Pick-Up	Dents
Non-Uniformed Agglomeration of Hardener Agents	Debond
Over- or Under-Cured	Edge Damage
Pills and Fuzz Balls	Erosion
Prepreg Variability	Excessive Ply Overlap
Surface Oxidation	Fastener Holes
Thermal Stresses	Fracture
Variation in Density	Holes and Penetration
Variation in Resin Fraction	Matrix Cracking and Crazing
Variation in Thickness	Mismatched Parts
Voids	Missing Plies
Wrong Materials	Ply Overlap or Gap
	Porosity
	Reworked Areas
	Surface Damage
	Surface Swelling
	Translaminar Cracks
	Variation in Thickness
	Voids
	Warping
	Wrong Material

edges, projectile impact and heat damage, the defects and damage in composite components and structures are classified in Table 2.4 for the typical location of in-service defects and damage only.

GENERALISATION OF NON-CONFORMITIES

The results of a detailed literature survey of defects types and associated stress state conditions that develop indicate that defects and damage in composite components and structures can be listed in terms of a developed common stress state. The common stress states are defined as either transverse (intralaminar) matrix cracks,

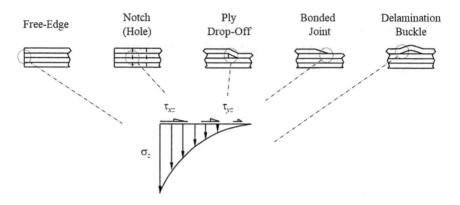

FIGURE 2.1   Sources of out-of-plane loads.

TABLE 2.4
Typical Location of In-Service Defects and Damage in Composite Materials

Geometric Discontinuities	Free Edges	Projectile Impact	Heat Damage
Corner/Edge Crack	Bearing Surface Damage	Crushing	Creep
Corner Radius Delamination	Delaminations	Cuts and Scratches	Matrix Cracking
Debond	Edge Damage	Delaminations	Matrix Crazing
Fastener Holes	Erosion	Dents	Surface Damage
Fibre Kinks	Fastener Holes	Debond	Surface Oxidation
Reworked Areas	Holes and Penetration	Fracture	Surface Swelling
	Moisture Pick-Up	Holes and Penetration	
		Surface Damage	
		Translaminar Cracks	

delaminations (interlaminar matrix cracks), fibre fracture (holes) or design variance. Under these four common stress state categories, the list of defects and damage in Table 2.1 are listed in Table 2.5.

REVIEW OF GENERALISED DEFECT TYPES

There are three basic stages of defect and damage development in composite materials. These stages of defect and damage development are matrix crack initiation, matrix crack growth and localisation of fibre breakage leading to the ultimate failure of the composite material. The fundamental failure mechanisms in composite materials are directly related to the generalised defects/damage types:

- transverse matrix cracking (intralaminar cracks)
- delaminations (interlaminar cracks)
- fibre fracture

TABLE 2.5
Generalised Defect and Damage Types in Composite Components and Structures

Delaminations	Matrix Cracks	Holes	Design Variance
Bearing Surface Damage	Bearing Surface Damage	Bearing Surface Damage	Creep
Blistering	Contamination	Crushing	Damaged Filaments
Contamination	Corner/Edge Crack	Cuts and Scratches	Dents
Corner/Edge Crack	Cracks	Fastener Holes	Erosion
Corner Radius	Edge Damage	Fibre Kinks	Excessive Ply Overlap
Delamination	Matrix Cracking	Fracture	Fibre Distribution
Delaminations	Matrix Crazing	Holes and Penetration	Variance
Debond	Porosity	Reworked Areas	Fibre Faults
Edge Damage	Translaminar Cracks	Surface Damage	Fibre Kinks
Fastener Holes	Voids		Fibre Misalignment
Fibre/Matrix Debond			Marcelled Fibres
Holes and Penetration			Miscollination
Pills and Fuzz Balls			Mismatched Parts
Surface Swelling			Missing Plies
			Moisture Pick-Up
			Non-Uniformed Agglomeration of Hardener Agents
			Overaged Prepreg
			Over- or Under-Cured
			Pills and Fuzz Balls
			Ply Underlap/Gap
			Prepreg Variability
			Surface Oxidation
			Thermal Stresses
			Variation in Density
			Variation in Resin Fraction
			Variation in Thickness
			Warping
			Wrong Materials

The failure mode of interfacial fracture between fibres and matrix can be treated as transverse matrix cracks but is an indication of poor process controls and/or fibre/resin selection. Interfacial fracture is very difficult to predict due to lack of measure of interfacial adhesion weakening.

Transverse Matrix Cracking

Transverse matrix cracks are the cracks within a ply of a laminate. These are the first cracks that appear in a laminate. They arise due to manufacturing process control issues (porosity is a transverse matrix crack) due to thermal induced stresses

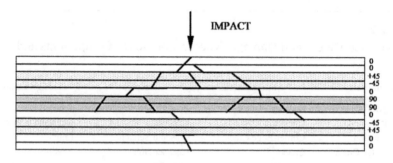

FIGURE 2.2   Matrix crack pyramid pattern from impact damage.

FIGURE 2.3   Transverse matrix cracks in a composite structure (highlighted).

on cool-down during fabrication from post-fabrication machining and light impact during in-service. Transverse matrix cracks will also initiate more significant fracture events. An example of the matrix crack generation from impact is depicted in Figure 2.2, and Figure 2.3 provides a photograph of actual transverse matrix cracks.

The occurrence of transverse matrix cracks is influenced by the following features of composite laminates:

- **Fibre volume fraction**. There is a higher potential of occurrence with lower fibre volume fraction.
- **Fibre packing uniformity**. Less uniformity (resin-rich areas) leads to easier creation of matrix cracks.
- **Degree of anisotropy** (E_f/E_m). Higher fibre stiffness to matrix stiffness reduces the load share in the matrix, reducing the transverse crack potential.
- **Environmental influences**. A hot/wet environment softens the matrix, reducing the formation of transverse cracks.

- **Toughness of the resin**. Greater toughness improves matrix resistance to fracture initiation.
- **Ply stacking configuration**. Well-distributed, well-oriented plies will provide mutual resistance to transverse crack formation.
- **Fabrics**. Woven composite fibre provides resistance to crack initiation. The lower the harness number of the woven fibres, the better resistance to transverse cracking in the matrix.

DELAMINATIONS

As the translaminar (intralaminar) matrix crack reaches the boundary of two adjacent plies of different orientation in unidirectional tape composites, or just another ply of fabric composites, the crack growth through the ply is halted. With the application of greater energy levels of loading and displacement, the translaminar crack changes direction into the plane of the laminate. The crack now propagates along the ply boundary interface through the resin-rich layer, as depicted in Figure 2.4.

The propagation of delaminations in graphite/epoxy composites is sensitive to:

- matrix toughness
- ply stacking sequence
- component shape and constraints
- delamination position
- delamination size
- load type and magnitude
- environmental effects

Generally, the presence of a delamination reduces the overall stiffness of a composite structure. This lowers the critical buckling load and can result in local laminate structural instability under compressive loading. The final failure of a delaminated structure is by:

- increased net section stresses
- out-of-plane bending
- asymmetric twisting

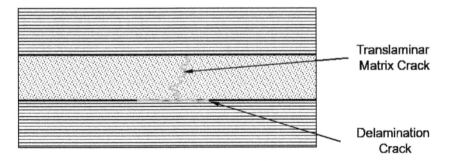

Translaminar
Matrix Crack

Delamination
Crack

FIGURE 2.4   Creation of the delamination from a translaminar matrix crack.

FIBRE FRACTURE

Fibre will fracture at much higher loads (forces) than the matrix material when under in-plane loads, particularly tensile loads. Axial compression loads in the fibres will be influenced by the elastic support of the matrix to prevent buckling and/or kinking of the fibres, but still at higher loads than the compression strength of the matrix. Interlaminar shear of the fibres is generally low in comparison with the fibre axial loads and considered in the weak interlaminar shear strength of the laminate. The fracture behaviour of laminates with holes under in-plane loading depends on the factors that follow:

- The notch-tip radius and the corresponding stress concentration developed at the notch tip at both the lamina and laminate levels
- The hole diameter and shape (circular, elliptical, etc.), and the relationship of the hole geometry with the panel geometry
- Which loading type is applied to the panel, such as axial tension/compression and interlaminar shear
- The laminate thickness effects, including changes in cross-sectional geometry with ply changes
- The ply orientation relative to the loading direction and the effect on the stress-concentration factor
- The laminate ply stacking sequence and the stacking of common plies together
- The difference between the ply material properties of the fibres and the matrix, in particular the Young's modulus and the associated global and local stiffness value

A phenomenon known as hole size effect can exist in composite laminates. Under in-plane loading, a larger hole causes a greater stress reduction than a smaller hole, although the maximum in-plane stress on the boundary of the hole is the same for all hole diameters. Moiré interferometry has indicated that high out-of-plane deflections or buckling increase the edge out-of-plane stresses. The fibres at the hole edge locally buckle and the damage propagates by shear crimping and delamination up to ultimate laminate failure. This buckling is more severe when the hole diameter-to-panel width ratio is small. The effect is primarily due to interlaminar stresses induced at the free edge of the hole, where cracks initiate and propagate in the matrix. The final fracture mode does, however, change from one point to another around the hole boundary, and in the end, the failure tends to be less than that predicted by classical lamination theory.

The fracture process of delaminations simply involves interlaminar cracking between two highly anisotropic fibre-reinforced plies. However, the fracture process is very complex because it depends on material and geometric discontinuity, and it appears to involve coupling effects of the three distinct modes of crack propagation: Modes I, II and III. Delaminations are initiated from either the free edge out-of-plane stress induction or growth of matrix cracks to ply interfaces (Figure 2.5). The progressive fracture behaviour of delaminations is shown in Figure 2.6 and

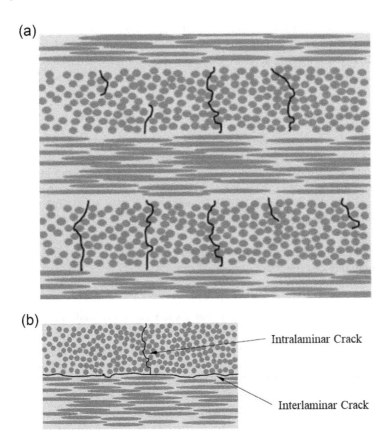

FIGURE 2.5   Matrix cracks and delamination initiation.

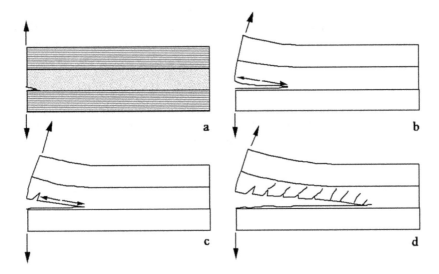

FIGURE 2.6   Delamination initiation and hackle formation.

illustrates the characteristics of matrix tearing and hackle formation. Due to the low interlaminar strength of untoughened graphite/epoxy composites, delaminations are easily initiated through impact damage or simple out-of-plane induced stresses under in-plane loading conditions.

CONCLUSIONS

Advanced polymeric composite materials are potentially prone to a large number of defects and damage types. These defects and damage types emanate from the constituent processing, component manufacture and fabrication, and in-service use (and abuse). Those defects and damage types that are due to in-service usage can be generalised into three categories for the purposes of similar stress state effects. The generalised categories are thus:

- Transverse matrix cracks (intralaminar cracks)
- Delaminations (interlaminar matrix cracks)
- Fibre fracture or holes

The failure modes of composite materials are numerous and are influenced by many factors. In multidirectional laminates, the prediction of failure modes is very difficult and is usually a combination of several unidirectional failure modes under the various loading spectrums and conditions. In the majority of failure cases, the failure mode is determined by post-mortem examination of the fracture surface.

What is typically seen, though, is a global self-similar damage progression. This damage progression tends to follow the three generalised categories of failure that are described in the following sub-paragraphs:

- Multiple matrix cracking will occur in the composite laminate. The multiple matrix cracking is often termed the Characteristic Damage State (CDS). CDS is the first stage of damage initiation and propagation. The damage pattern at the CDS stage is random and scattered over the laminate. In some very specific damaging conditions, the CDS will be very localised, for example at impact zones.
- The second stage of damage progression is the initiation of delaminations or significant transverse matrix crack localisation. At this point in the damage progression, the damage density increases in size at preferred sites of the composite laminate. Such localisation of the CDS is at laminate free edges or internal ply interfaces.
- The final fracture of the damage progression is often multi-moded with severe and extensive matrix cracking, but fibre fracture will be the controlling factor in the composite laminate.

Composite fracture behaviour for untoughened graphite/epoxy laminates in the presence of holes and delaminations is usually governed by the interlaminar structural integrity of the laminate.

3 Damage Assessment

INTRODUCTION

Non-Destructive Inspection (NDI) methods are employed in the repair process of composite and adhesively bonded structures in three ways:

- Damage location
- Damage evaluation, i.e. type, size, shape and internal position of the damage
- Post-repair quality assurance

The first and most important activity in a repair process is to identify the defect or damage. Assessment of the damage is usually initially achieved by visual inspection during a structure walk-around survey. This initial visual inspection localises the damaged area, and then a more sensitive NDI method is used to map the extent of any external/internal damage. Detailed NDI is very important when dealing with composite and adhesively bonded structures because of the concern and insidious nature of composite damage which can be hidden internally, see Figures 3.1 and 3.2, for example.

The types of NDI methods typically available for inspecting composite and adhesively bonded structures are as follows:

- Visual inspection includes methods with optical magnification and defect enhancement. Visual inspection is simply using the inspector's own eyes to identify the location of surface damage.
- Acoustic inspection methods can identify changes in sound emission, which varies with the presence of external and internal damage.
- Ultrasonic inspection methods, such as A-Scan and C-Scan, that use the travel of sound energy through the part and the reflective/transmission can identify variations due to the presence of any internal damage.
- Thermography inspection methods track the changes of heat transmissivity through a composite part and the variation when internal damage is present.
- Interferometry inspection methods and all of the optical methods that analyse the light reflectivity changes with structure response due to the presence of internal and external damage.
- Radiography inspection methods are used to penetrate the component with energy-stimulated particles to reveal internal anomalies in the composite and adhesively bonded structures.

- Microwave inspection methods are used for the assessment of moisture contamination primarily of composite and adhesively bonded structures.
- Material property changes inspection methods, such as stiffness and dielectric, are used to identify property variations due to the presence of the defect.

Each of these NDI methods used in damage and defect identification in composite and adhesively bonded structures are briefly discussed in the following sections.

METHODS OF COMPOSITE NDI

The majority of NDI methods available for composite laminates and adhesively bonded structures have been used successfully with metallic structures. However, when used on composite and adhesively bonded structures some changes to operating parameters are necessary, and the interpretation of the NDI result requires training and a different understanding in many instances. However, due to the diversity of defect types likely to be found in composites and adhesively bonded structures,

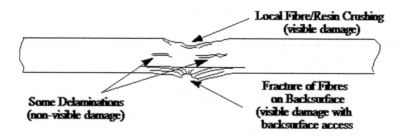

FIGURE 3.1 Internally hidden damage with external visible damage.

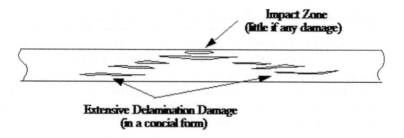

FIGURE 3.2 Barely visible impact damage (BVID).

several methods may be required to fully detail the defect or damage. This situation can mean the acquisition of a number of different types of expensive NDI equipment and a requirement to have highly trained NDI inspectors/assessors.

The following discusses in general terms the most commonly used NDI techniques for composite and adhesively bonded structures. At the end of this section, a number of more advanced systems are outlined.

VISUAL INSPECTION

Apart from simply using one's eye, which only identifies obvious surface defects such as that shown in Figure 3.1, simple magnification can identify quite small surface defects. To improve defect or matrix crack visual clarity, enhancement with a dye penetrant can be used; however, this may cause long-term issues with the dye staying in the surface cracks. Some internal defects can also be found using borescope methods, but access is still required, i.e. through a fastener hole. Bondline visual inspection will provide some assessment of the resin flow. The typical resin flows from a bonded joint edge are shown in Figure 3.3.

- Basic visual methods:
  - are inexpensive
  - are relatively simple
  - require relatively low skill levels
  - need the surface cleaned
  - are only suitable for surface defects

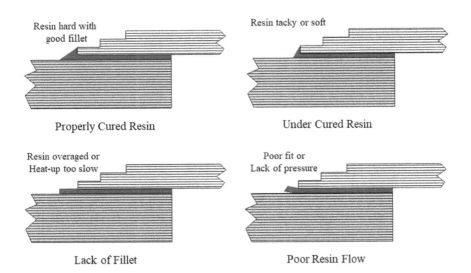

FIGURE 3.3 Bonded joint edge resin flows.

- Dye penetrant methods:
  - contaminate the surface to be inspected
  - are only suitable for surface defects
  - require pre-cleaning and post-cleaning of the part surface
  - are portable
  - are relatively simple to apply
  - require some level of operator skill
- The hot water leak test:
  - requires a large water tank to be heated to 65°C (150°F)
  - will check the water tight integrity of sealed structures (sandwich structures)
  - **requires** radiography to disclose water entrapment if the part fails the hot water leak test

ACOUSTIC METHODS

All of the acoustic emission methods require the operator or acoustic equipment listening to crack growth via changes in sound from a light impact or elastic wave energy.

- Coin-Tap (Hammer) Method (Figure 3.4):
  - is a simple method to use
  - is suitable to near surface defects (shallow) and is ideal for thin skinned sandwich structures
  - is dependent on the part geometry
  - requires two-sided access for thicker panels and sandwich structures
  - is very portable
  - requires the operator to have a good ear
- Acoustic Emission (Figure 3.5):
  - will require experienced operators
  - has a complex output
  - is semi-portable and recordable
  - is sensitive to very small changes in a defect

ULTRASONIC METHODS

The implementation of ultrasonic inspection can range from inexpensive to quite costly in terms of the equipment requirements. The methods can simply provide details of depth and size of the non-conformity, or full details of the sub-surface defect topography. There are two principle methods, pulse-echo (A-scan) or through transmission (C-scan). Both methods measure change in sound attenuation (amplitude loss). C-Scan methods are illustrated in Figures 3.6 through 3.8, and the typical results are shown in Figure 3.9.

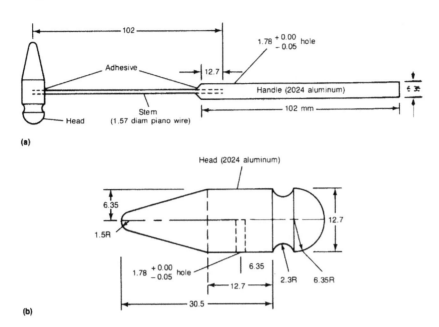

FIGURE 3.4   Tap-test hammer (dimensions in mm).

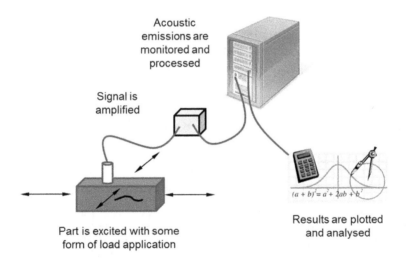

FIGURE 3.5   Acoustic emission detection of a composite panel.

- Pulse-Echo method:
    - will display the amplitude of the return signal versus time of fight

- requires a coupling agent to allow the signal to penetrate the specimen
- provides topographic information on defect type, size, location and depth
- is reasonably sensitive to small defects
- requires a standard specimen as a benchmark to judge defective material
- requires experienced operators of the equipment and results interpretation
- is portable, and the output is recordable
- will require pre-cleaning and post-cleaning of the part prior to examination
- Through transmission (immersion) is generally the same as pulse-echo, except that it:
  - is automated and therefore faster than through transmission.
  - only provides definitions of defect size and location
  - provides full coverage of the component
  - required double-sided access to the part being inspected
  - is for internal defects only

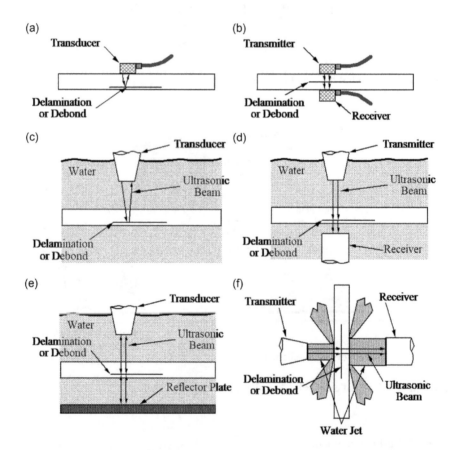

FIGURE 3.6   Ultrasonic inspection techniques. (a) Contact pulse-echo; (b) Contact through transmission; (c) Immersion pulse-echo; (d) Immersion through transmission; (e) Immersion reflection; (f) Water jet through transmission.

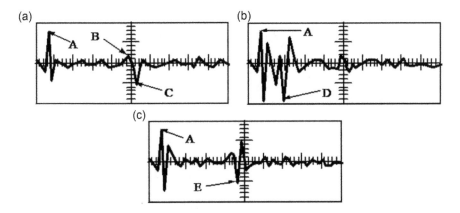

FIGURE 3.7 Representative ultrasonic pulse-echo results of a graphite/epoxy composite skin and honeycomb core. (a) A good bonded sample; (b) Front skin delamination; (c) Debond between skin and core.

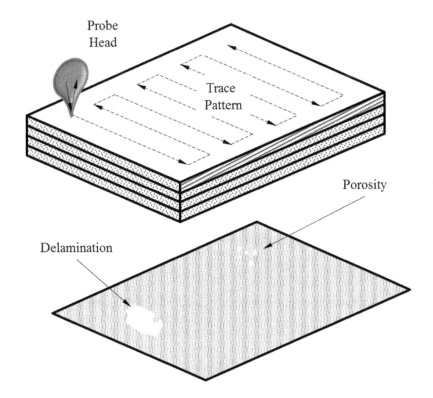

FIGURE 3.8 Schematic of the ultrasonic C-scan.

FIGURE 3.9   Ultrasonic C-scan results of a graphite/epoxy laminate.

THERMOGRAPHY

Thermography as an NDI technique that measures the response of a structure to thermal energy dissipation. The application of thermal energy can be passive or active. Passive thermography is when heat is applied to the structure (i.e. a flash lamp), and the thermal wave can be visualised on an infrared camera. Whereas active thermography is the structure is excited and the variations in the stress produce changes in the thermal condition of the structure at a local level. These thermal changes are also captured on in infrared camera. Thermography is very applicable to composite materials and internal defects. When defects are present in the form of non-contacting surfaces within a composite laminate, then the rate at which thermal energy dissipates (passive) is reduced, or thermal energy increases (active) can be measured. Both passive and active thermography methods:

- require standards to verify results
- are a very portable and recordable system
- require **experienced** operators and assessors
- can be geometry dependent

Typical results are shown in Figure 3.10.

INTERFEROMETRY

The use of light and its reflective properties to identify defects can be described as optical interferometry NDI. There are three basic methods optical interferometry NDI:

- Moiré
- Holography see Figures 3.11 and 3.12, and
- Shearography, see Figure 3.13.

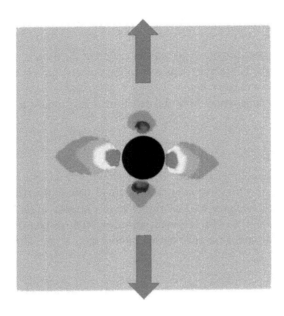

FIGURE 3.10   Schematic of thermography results.

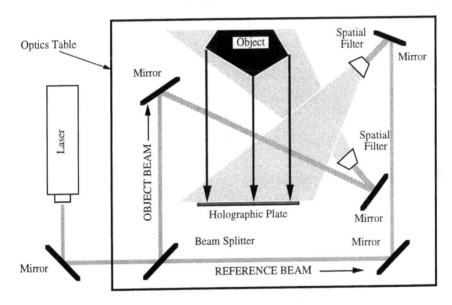

FIGURE 3.11   Conventional holography procedure.

Aspects to consider when using optical interferometric methods are:

• that the equipment can be expensive
• the need for skilled (qualified) operators and results interpreters

- that the methods are generally not portable (but modernisation is improving in this area)
- that the methods provide a permanent record of the defects
- that the methods are very sensitive to the defects' presence on structural performance and behaviour
- that the methods show how the structure reacts under a given loading condition

RADIOGRAPHY

The principle of radiography as an NDI technique is to uncover internal non-conformities by how much radiation is absorbed and how much radiation is allowed to pass through the component. This principle of radiation as an NDI technique is illustrated in Figure 3.14. The two main radiation NDI methods used in composite sandwich structures are x-ray and neutron radiography, x-ray radiography being the

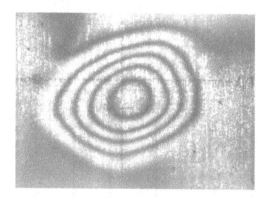

FIGURE 3.12   Double-exposed holography showing defects.

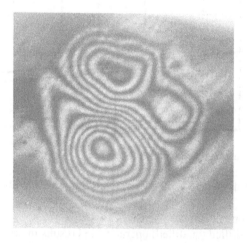

FIGURE 3.13   Defect evaluation using shearography.

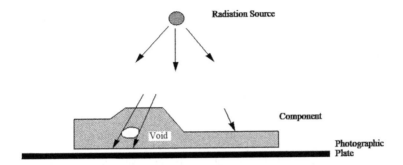

FIGURE 3.14 Principle of radiography.

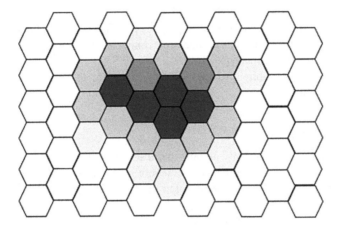

FIGURE 3.15 X-ray radiograph of honeycomb core with water entrapment.

most common. Typical output results from both x-ray (Figure 3.15) and neutron radiography that are typical of sandwich structure core assessment.

X-Ray Radiography

The x-ray radiography method and approach can be summarised as follows:

- The results of x-ray radiography are easy to interpret as the radiographic images are visually understandable.
- X-ray radiography is excellent for honeycomb sandwich panel inspection to identify issues with the core itself.
- The radiographic images provide a permanent record of the inspection and allow for future comparative images.
- The basic x-ray radiography equipment is relatively expensive, although the costs are reducing with digital technology being implemented in the traditional x-ray radiography process.
- X-ray radiography is now easily portable with digital radiography process and equipment on the market.

- However, irrespective of the form of x-ray radiography, it still requires a set of **strict safety** procedures to be implemented.
- X-ray radiography will require experienced operators.
- The application of dye penetrant will provide enhanced x-ray radiography results. However, the use of dyes will likely cause problems with bonded repair durability and are thus discouraged in composite structure and sandwich structure work.

Neutron Radiography

The neutron radiography method and approach can be summarised as follows:

- Neutron radiography is very expensive, in particular the neutron particle machinery and safety shielding facilities.
- As an NDI process, neutron radiography is similar to x-ray radiography.
- Neutron radiography is excellent for composite materials and moisture entrapment in sandwich panels.
- Neutron radiography provides better resolution of results over x-ray radiography.

MICROWAVE

Microwave NDI is used mainly on **non-metallic** materials to determine the degree of moisture content through the measurement of microwave absorption. The microwave NDI method requires the following to support NDI outcomes:

- The test region normally requires two-sided access; however, new approaches are allowing one-sided access.
- Any metal components will require shielding.
- The operator will need to follow strict **safety procedures**.
- The component must be thoroughly clean.

MATERIAL PROPERTY CHANGES

There are two major material property change NDI methods used with composite structures. They are the measurement of dielectric and stiffness properties. Dielectric measures the degree of cure of an adhesive or composite resin by reading electrical inductance. Stiffness changes are determined from mechanical testing of load versus deflection, and a comparison is made against theoretical analysis results. Both methods have their limitations.

IN SITU METHODS

A significant amount of research effort and some regular application of in situ NDI is being used with composites. Such in situ NDI methods allow for the identification of concerning damage and defects as they happen. Such methods of in situ NDI are considered as:

- **Optical fibres**. The use of embedded optical fibres allows the transmission of light, and any interruption to this transmission will indicate an issue with the optical fibre integrity. This loss of optic integrity is assumed to be caused by damage to the structure where the optical fibre is positioned. The optical fibre receiver can then identify the exact location of the structural concern. Hence, the position of the optical fibre network within the structure is in areas of most structural concern.
- **Piezo-electric sensors**. Like optical fibres, the position of a piezo-electric sensor will be at locations of critical structural integrity concern. However, where optical fibres provide a line of sensory ability, the piezo-electric sensor will only give the point location of an issue. But, the piezo-electric sensor will provide an indication of the local stress state and the influence of stress state due to regional loading effects.
- **Embedded strain gauges**. Similar to piezo-electric sensors, the embedded strain gauge will provide a local stress state and reflect changes to the local stress condition that can be caused by regional loading effects. Damage to an area within the structure the influences the stress state where the embedded strain gauge is located could cause a change in the stress state at the strain gauge and thus be tagged as an issue within the structure.

With any embedded sensor, once there is damage to the sensor, then repair of the sensor can be a major issue, and more so than the repair of the structure. Currently embedded sensors have been used in only a few applications and are still under research development. The major reason for this limited use of embedded sensors is that their presence in the composite structure acts as an inclusion and are typically the point of initial and premature failure of the composite structure.

APPLICATION OF NDI METHODS

For the successful application of any NDI method, a selection process of the most suitable method must be in place, and the required skilled personnel must be known. The selection process for the most suitable NDI method is typically based on experience, but a formal process can be developed that would apply to unique situations. A combination of equipment availability and personnel who have been trained on and have experience with a particular NDI method is essential for the correct location and characterisation of composite defects and damage.

NDI SELECTION PROCESS

The NDI selection process is based on the following four criteria:

- The first criterion is to understand the configuration of the component to be investigated, the orientation of the composite laminate, and the materials it is made from. Such knowledge allows for better equipment selection and interpretation of the results from an NDI survey.
- An idea of the type, size and location of the defect/damage to be inspected significantly aids in detection and assessment.

- Relative accessibility to the assessment area of the component may restrict the NDI process to be used. Limited space and position of the component for adequate NDI assessment can place significant challenges on obtaining credible outcomes of the survey.
- The availability of NDI equipment, accessories, consumables, space to undertake the NDI survey and appropriately skilled NDI technicians will enhance the outcomes of the NDI survey. A list of NDI equipment availability with accompanying accessories and consumables, and the NDI technician level to operate such NDI equipment should be available to engineering and technical managers.

The ability and success of the various NDI methods to find various defects in composite structures are listed in Table 3.1.

NDI PERSONNEL

The NDI operator and assessor must be:

- conversant with several different inspection techniques
- able to set up the equipment and effectively modify the standard diagnostic arrangements to suit the target
- skilled to interpret the resulting NDI information
- knowledgeable of safety standards and procedures
- able to comply with MIL-STD-410 or its equivalent

IMPORTANT REQUIREMENTS

For NDI to be successful in detecting the extent of damage in composite and adhesively bonded structures and components, three requirements must be satisfied. They are as follows:

EQUIPMENT AND FACILITIES

The suitable NDI equipment and facilities, including personnel safety and environmental health procedures, must be available, calibrated and in good working order.

TRAINED OPERATORS

The operators of NDI equipment must be adequately trained and experienced to ensure that the results from any damage assessment survey are both accurate and reliable.

COMPARATIVE SPECIMENS

Any NDI technique is comparative in nature, that is the results of an assessment survey are usually compared with a good or like damaged specimen. This is particularly important when calibrating NDI equipment.

TABLE 3.1

NDI Methods versus Defect Type Assessment Success

Category	Defect	Visual	Dye Penetrant	Coin/Hammer Tap Testing	Bondoscope Ultrasonics	Pulse-Echo Ultrasonics	Through-Transmission Ultrasonics	X-Ray Radiography	Dielectric Change Detection	Thermography	Optical Interferometry	Microwave Absorption	Neutron Radiography	Frequency Response Changes
Laminate	Delaminations	1,2	1	√	√	√	√	3		√	√			√
	Macrocracks	1,2	2	√	√			3		√	√			
	Fibre Fracture							√		2,3	2,3			
	Interfacial Cracks									2,3	√			
	Microcracks		1	2	2					√	√			
	Porosity	1		2	2	√	√	√		2	√			
	Inclusions	1			2	2	2	√		√	√			√
	Heat Damage	1		2	2				2	2				
	Moisture							2	√	2		√	√	
	Voids				2	√	√	√		√	√			
	Surface Protrusions	√								√	√			
	Wrinkles	√								√	√			
	Improper Cure								√	2	2		√	
Bondline	Debonds	1,2	1	√	√	√	√	√		√	√			√
	Weak Bonds									2	√			
	Cracks	1,2	1	2	2	2	2	3		√	√			
	Voids			√	√	√	√	√		√	√			√
	Moisture							√	√	2		√	√	
	Inclusions			2	2	2	2	√		√	√			√
	Porosity				2	√	√	√		√	√			
	Lack of Adhesive			√	√	√	√	√		√	√			
Sandwich Panels	Blown Core			√	√	√		√					√	√
	Condensed Core			2	2		2	√					√	
	Crushed Cure			2	2		2	√					√	
	Distorted Core							√					√	
	Cut Core			√	√		√	√					√	
	Missing Core			2	2	2	2	√					√	√
	Node Debond							√		2	√			
	Water in Core			2	2		2	√				√	√	
	Debonds			√	√	√	√	√		√	√			√
	Voids			2	2	√	√	√		√	√			
	Core Filler Cracks			2		2		√	3	2	2			
	Lack of Filler			2	2	2		√	√	2	2		√	

Notes: 1. Open to surface.

2. Unreliable detection.

3. Orientation dependent.

4 Damage Stress Analysis

INTRODUCTION

Damage stress analysis of composite structures is based on the stress state around the damaged area. The stress state analysis will include the influential aspects, such as stress concentration development, environmental degradation of mechanical properties, damage location, damage orientation and damage complexity. The estimated stress state is then compared with the component's ultimate strength or the first ply failure strength or the design limit stress state to determine the severity of the damage. This approach to damage stress state estimation is typically conservative, especially when considering the current material load performance requirements. However, the damage stress state estimation will provide the engineer with an indication of what type of repair scheme needs to be designed. For example, there is a cosmetic repair scheme for environmental exclusion, or a semi-structure repair scheme for damage tolerance restoration, or a structural repair scheme for restoration of structural integrity, if needed.

STRUCTURALLY SIGNIFICANT DAMAGE TYPES

In the evaluation of defect and damage types' criticality, we can generalise the defects and damage into three categories. These categories are identified as matrix cracks (including heat damage), delaminations and holes (fractured fibres)(Figure 4.1), and are the main structural performance degradation damage concerns in composite structures.

These damage types are briefly redefined in the following paragraphs.

MATRIX CRACKS

Matrix cracks in composite laminates are confined to those cracks within a ply or lamina. Matrix cracks are often termed intralaminar cracks. Intralaminar cracks tend to be transverse to the fibre direction and terminate at the ply boundaries (Figure 4.2). Local stiffness loss is usually attributed to intralaminar matrix cracks on a composite laminated structure. Heat damage is a common cause of intralaminar matrix cracks.

DELAMINATIONS

Delaminations are also a form of matrix cracking, but this time, these matrix cracks lie in the plane of the laminate and between plies. Delaminations are also termed interlaminar matrix cracks and normally grow from intralaminar matrix cracks when the cracks terminate at a ply boundary (Figure 4.3). Structural instability

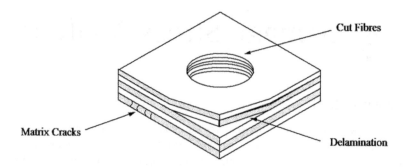

FIGURE 4.1   Principal damage types in composite laminates.

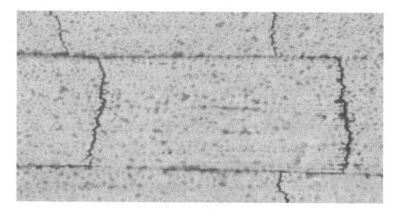

FIGURE 4.2   Intralaminar matrix cracks.

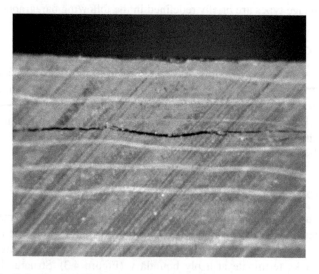

FIGURE 4.3   Delaminations (interlaminar matrix crack).

under compressive loading is of the greatest concern for the loss of structural performance of a composite laminated structure from the presence of delaminations.
Delaminations can be found in several locations within a composite laminate,
including internally, along a free edge or on a corner of the composite laminate
(see Figure 4.4). Delaminations can appear as a single delamination or as multiple
delaminations, as shown in Figure 4.5.

HOLES

Holes and cut fibres in composite laminates are areas of missing, discontinuous or
fractured fibres and matrix constituent materials. Holes are the clear absence of
adjoining constituent materials but must include fibre breakage, as seen in Figure 4.6.
Holes in composite laminated structures are either entirely through-the-thickness
or as a partial penetration. A partial fractured surface is illustrated in Figure 4.7.
With the presence of holes (fractured fibres) in a composite structure, there is an

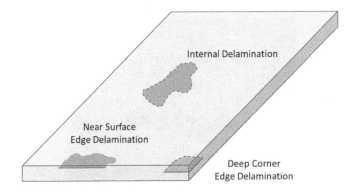

FIGURE 4.4   Typical locations of a delamination in a composite laminated structure.

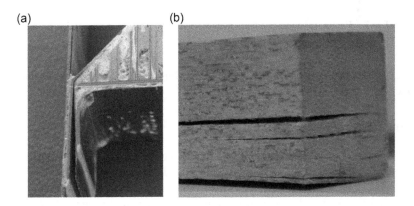

FIGURE 4.5   Single and multiple delaminations. (a) Single delamination; (b) Multiple
delaminations.

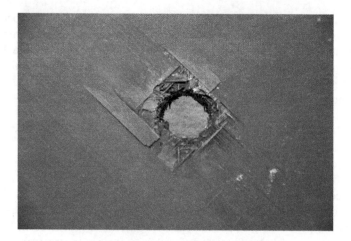

FIGURE 4.6   Hole in a composite laminated structure caused by impact.

FIGURE 4.7   Partial through-the-thickness hole in a composite laminated sandwich structure (one facing of the composite sandwich structure is penetrated).

induced stress concentration and a corresponding reduction in strength under in-plane loading.

STRESS STATE AND FAILURE CRITERIA

With composite materials, the normal stress state is quite complex, particularly at any edges such as free edges, holes or defects and damage. As such, the stress state at any free edge is truly three-dimensional. Thus, in addition to the in-plane stresses (σ_x, σ_y and τ_{xy}), the interlaminar stress components σ_z, τ_{xz} and τ_{yz} are induced into the laminate. These interlaminar stresses can be difficult to estimate and/or measure. Although the interlaminar stress components are necessary in the complete stress state analysis, we can easily justify using only in-plane stress analysis based on the knowledge that the current design allowable strains provide sufficient damage tolerance, and thus we can ignore (but not forget) the interlaminar stresses.

FAILURE CRITERIA

Once the stress state of the damaged area is known (calculated or estimated), we need to evaluate it against the composite material properties. This comparison is usually done via a suitable failure criterion. The assessment of structural integrity, by way of a failure criterion, the design and induced stress states are compared with the known composite material properties such that one of the following conditions applies:

- Failure Criterion < 1   ......   no failure exists in the structure
- Failure Criterion = 1   ......   failure is imminent in the structure
- Failure Criterion > 1   ......   the structure has failed in some manner

There are two types of failure criteria methodologies used for the assessment of structural integrity changes. The first failure criterion methodology uses linear elastic fracture mechanics (LEFM), and the second failure criterion is based on stress analysis at a particular point in the structure. Both are briefly described in the following paragraphs.

Linear Elastic Fracture Mechanics (LEFM)

LEFM is a relatively new stress analysis technique that has provided significant advances in stress analysis of metallic structures, particularly under fatigue loading. However, the application of LEFM has not shown itself to be specifically relevant to composite materials but has shown some applicability to adhesively bonded joints. The main problem with applying LEFM to composite materials tends to be associated with the local lack of material homogeneity. Metal materials are isotropic, and the evaluation of the LEFM stress state can accommodate this way the similar crack growth approach. Also, composite materials, being uniquely anisotropic at the ply level, restrict intralaminar crack growth when the fibres are encountered at the ply interface, and cracks in the composite tend to be in the resin-rich interlaminar zones or ply boundaries.

The two LEFM methods used in structural analysis are:

1. Strain Energy Release Rate (G)
2. Strain Energy Density (S)

The respective failure criterion for both the LEFM methods can be expressed as:

1. Strain Energy Release Rate,

$$\left(\frac{G_{\mathrm{I}}}{G_{\mathrm{Ic}}}\right)^{m} + \left(\frac{G_{\mathrm{II}}}{G_{\mathrm{IIc}}}\right)^{n} + \left(\frac{G_{\mathrm{III}}}{G_{\mathrm{IIIc}}}\right)^{p} \leq 1$$

where m, n, p and G_{Ic} are determined experimentally

2. Strain Energy Density,

$$S_{min} \leq S_C$$

where S_C is a material parameter determined experimentally.

Crack growth will occur when the measure values equate to material property or the critical value.

Point Stress Analysis

The stress state at a particular point in the composite structure is evaluated against the composite material strength properties directly through a failure criteria mathematical expression. The failure criterion can be a relatively simple direct relationship of the stress in one direction to its corresponding strength value or the more complex methods that provide interaction of the stress state. Several of the more commonly used point stress type failure criterion in composite structures are discussed next (note that several other failure criteria exist; see the Failure Criteria in Fibre-Reinforced-Polymer Composites: The World-Wide Failure Exercise).

1. **Maximum Stress Failure Criterion**. The Maximum Stress Failure Criterion has been used extensively by the aerospace industry in the early years of composite structure design. It is still used as the primary method of evaluating adhesively bonded joints. While it is very simple to use, there are limitations in over-estimating the shear performance. The Maximum Stress Failure Criterion is expressed as

$$\frac{\sigma_x}{X} < 1, \quad \frac{\sigma_y}{Y} < 1, \quad \frac{\sigma_z}{Z} < 1$$

   where

   σ_x, σ_y and σ_z are the applied stresses in the x-axis, y-axis and z-axis directions.
   X = The x-axis ultimate strength
   Y = The y-axis ultimate strength
   Z = The z-axis ultimate strength

2. **Maximum Strain Failure Criterion**. The Maximum Strain Failure Criterion is a derived version of the Maximum Stress Failure Criteria, but this criterion takes into consideration the Poisson's Ratio effect. The Maximum Strain Failure Criterion is expressed as

$$\varepsilon_x < \frac{X}{E_x}, \quad \varepsilon_y < \frac{Y}{E_y}, \quad \varepsilon_z < \frac{Z}{E_z}, \quad \text{and}$$

$$\gamma_{xy} < \frac{S}{G_{xy}}, \quad \gamma_{xz} < \frac{R}{G_{xz}}, \quad \gamma_{yz} < \frac{T}{G_{yz}}$$

where

ε_x, ε_y and ε_z are the applied axial strains in the x-axis, y-axis and z-axis,
γ_{xy}, γ_{yz} and γ_{xz} are the applied shear strains in the xy-plane, yz-plane and xz-plane,
E_x, E_y and E_z are the orthogonal Young's moduli in the x-axis, y-axis and z-axis, and
G_{xy}, G_{yz} and G_{xz} are the shear moduli in the xy-plane, yz-plane and xz-plane.

3. **Tsai-Hill Failure Criterion**. Based on a previous modification of the von Mises failure criteria, the Tsai-Hill Failure Criterion considers the laminate transverse isotropy in its derivation. The Tsai-Hill Failure Criterion provides an enhanced assessment of the stress state through an interaction condition. In the 1-2 plane of the laminate, the individual plies are assessed through the following expression:

$$\left[\frac{\sigma_x}{X}\right]^2 - \frac{\sigma_x\sigma_y}{X^2} + \left[\frac{\sigma_y}{Y}\right]^2 + \left[\frac{\tau_{xy}}{S}\right]^2 < 1$$

where

σ_x, σ_y and τ_{xy} are the x-y in-plane axial stresses and shear stress in each ply of the laminate.
X = The x-axis ply ultimate strength
Y = The y-axis ply ultimate strength
S = The in-plane ply ultimate shear strength

4. **Hoffman Failure Criterion**. The assessment of all six components of the stress state (the orthogonal axial stress and shear stresses) can be achieved through the Hoffman failure criterion:

$$C_1\left(\sigma_y - \sigma_z\right)^2 + C_2\left(\sigma_z - \sigma_x\right)^2 + C_3\left(\sigma_x - \sigma_y\right)^2$$
$$+ C_4\sigma_x + C_5\sigma_y + C_6\sigma_z + C_7\sigma_{yz} + C_8\sigma_{xz} + C_9\sigma_{xy} < 1$$

where

$$C_1 = 0.5\left[\frac{1}{Y_tY_c} + \frac{1}{Z_tZ_c} - \frac{1}{X_tX_c}\right]$$

$$C_2 = 0.5\left[\frac{1}{X_tX_c} + \frac{1}{Z_tZ_c} - \frac{1}{Y_tY_c}\right]$$

$$C_3 = 0.5\left[\frac{1}{X_tX_c} + \frac{1}{Y_tY_c} - \frac{1}{Z_tZ_c}\right], \quad \text{and}$$

$$C_4 = \frac{1}{X_t} - \frac{1}{X_c} \quad C_5 = \frac{1}{Y_t} - \frac{1}{Y_c} \quad C_6 = \frac{1}{Z_t} - \frac{1}{Z_c}$$

$$C_7 = \frac{1}{T^2} \quad C_8 = \frac{1}{R^2} \quad C_9 = \frac{1}{S^2}$$

Subscripts t and c are the tensile and compression strength.

R and T are the interlaminar shear strength of the x-z plane and y-z plane.

5. **Quadratic Polynomial Failure Criterion**. The most commonly engaged failure criterion in the aviation industry for composite materials is the Quadratic Polynomial Failure Criterion. The Quadratic Polynomial Failure Criterion is a generalisation of the tensor failure criteria used in isotropic materials. The Quadratic Polynomial Failure Criterion is also known as the Tsai-Wu Failure Criterion. The Quadratic Polynomial Failure Criterion in general three-dimensional terms is expressed as

$$F(\sigma) = F_i \sigma_i + F_{ij} \sigma_i \sigma_j < 1 \quad \dots i, j = 1, \dots, 6$$

where

$$F_1 = \frac{1}{X_t} - \frac{1}{X_c} \quad F_2 = \frac{1}{Y_t} - \frac{1}{Y_c} \quad F_3 = \frac{1}{Z_t} - \frac{1}{Z_c}$$

$$F_4 = 0.0 \quad F_5 = 0.0 \quad F_6 = 0.0$$

$$F_{11} = \frac{1}{X_t X_c} \quad F_{22} = \frac{1}{Y_t Y_c} \quad F_{33} = \frac{1}{Z_t Z_c}$$

$$F_{44} = \frac{1}{T^2} \quad F_{55} = \frac{1}{R^2} \quad F_{66} = \frac{1}{S^2}$$

$$F_{ij} = F_{ij}' \sqrt{F_{ii} F_{jj}} \quad i, j = 1, 2, 3 \text{ and } i \neq j \quad F_{ij}' = -0.5$$

6. **Yamada Failure Criterion**. If all the matrix strength parameters of the Quadratic Polynomial Failure Criterion are reduced to zero, i.e. the matrix has completely cracked, then the Yamada Failure Criterion represents the last ply failure condition. The Yamada Failure Criterion is expressed as

$$\left[\frac{\sigma_x}{X} \right]^2 + \left[\frac{\tau_{xy}}{S} \right]^2 < 1$$

A couple of the failure criteria are mapped out in Figure 4.8 for the in-plane x-y plane.

Several other failure criteria have been and are being used, such as the LARC, Puck, Hashin, SIFT, etc. A detailed, descriptive list of many failure criteria for composite materials can be found in Failure Criteria in Fibre-Reinforced-Polymer Composites: The World-Wide Failure Exercise.

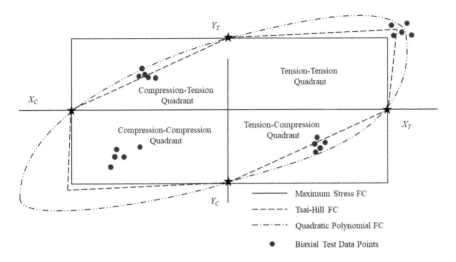

FIGURE 4.8 Several failure criterion mapped on the *x-y* plane.

OUT-OF-PLANE STRESSES

In laminated composites structures, the strengths (normal and shear) through-the-thickness, or *z*-axis direction, are notoriously weak when compared with the fibre dominated strengths. Due to a Poisson's ratio mismatch effect at the free edge of a composite laminate, the in-plane stresses induce the out-of-plane stresses (Pagano and Pipes). Thus, at the free edge, the stress state in the laminate is three-dimensional and is often the cause of premature failure through intralaminar and interlaminar crack initiation and propagation. The induced out-of-plane stresses arise due to component geometries, as shown in Figure 4.9. The three out-of-plane or interlaminar stress components σ_z, τ_{xz} and τ_{yz} produce an edge stress distribution at a ply interface typical of that shown in Figure 4.10. The magnitude of these interlaminar

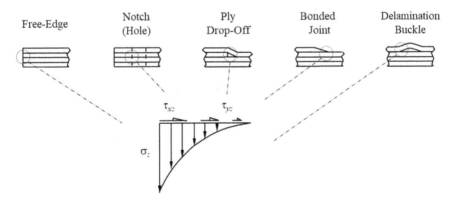

FIGURE 4.9 Out-of-plane stresses induced in composites laminates.

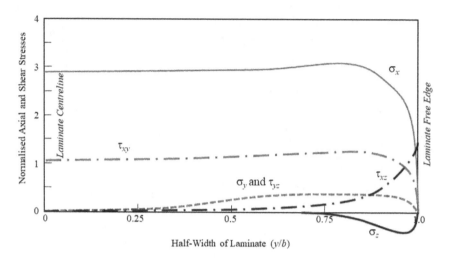

FIGURE 4.10   Interlaminar stress distribution at a free edge.

stresses decays rapidly inboard of the free edge and within approximately the thickness of the laminate. However, generally, the peak values of the interlaminar stresses are at or near the free edge. Likewise, we find that the axial stress parallel with the free edge decays to zero at the free edge. The laminate stacking sequence has a significant influence on the magnitude of the interlaminar stresses, and a good dispersion of plies is known to significantly improve interlaminar strength and ultimately increase the in-plane laminate strength.

THE REPAIR PROCESS

A review of the repair process (Table 4.1) suggests that following damage location and assessment (NDI), the stress state of the damaged area is to be conducted. Whilst the damage stress analysis can be difficult and complex, it still must be done to ensure that the correct repair approach is adopted. There have been too many times when major structural repair has been done where a simple repair scheme would have

TABLE 4.1

Repair Development Methodology

1. Locate the damaged area
2. Assess the extent of damage
3. **EVALUATE THE DAMAGED AREA STRESS STATE**
4. Design the repair scheme
5. Fabricate and prepare the repair scheme
6. Apply the repair scheme
7. Conduct post-repair quality checks
8. Monitor the repair region

sufficed. We also know that on many an occasion the structural repair has caused a great reduction in the structural integrity of the composite structure compared to just leaving the damage in place. Damage stress analysis is then followed by the design of the repair scheme based on the requirements of what needs to be achieved with the repair scheme. Therefore, to ensure that the most appropriate and cost-effective repair scheme is designed and installed, the stress state of the damage area must be evaluated. Unlike conventional structural materials and construction methods, the severity of the damage to composite structures is not always obvious.

MATERIAL PROPERTIES DETERMINATION

With both the LEFM and point stress analysis methods, the damage stress state in the composite structure is compared with the composite material properties, either at the global or local (ply) levels. For the most part, the damaged stress state is evaluated at the laminate level and thus requires an estimation of the laminate structural strength of the composite component. The laminate strength is calculated from the individual ply properties that make up a composite materials property database. In the point stress analysis method, these material ply strength values are the mutual perpendicular or normal strength values (X, Y and Z), and the corresponding ply shear strengths (S, R and T). The material strength properties are determined experimentally or estimated from known constituent material property by micromechanics. Several assumptions are made to reduce the level of testing and/or estimation of the 3D ply properties, such as the transverse isotropy assumption, where $Y = Z$ and $S = R$ for unidirectional composite plies. The ply properties are then used to evaluate the laminate properties through stress analysis processes like classical laminate plate theory and using one of the failure criteria discussed earlier.

Likewise, the two LEFM methods discussed earlier have critical values G_c and S_c, which are determined experimentally. However, for composite material tests such LEFM, property values have not shown true material property consistency. As a result of these inconsistencies with development of the LEFM material property value, LEFM material property databases availability is very limited. Hence, in this book, LEFM approaches in the development of stress states at damaged locations in composite structures will not be considered any further.

STRESS ANALYSIS OF DAMAGE TYPES

We now focus our discussions on the methods of analysing the stress state or structural performance changes due to the principal (generalised) damage types identified in Chapter 2. We shall develop the loss of structural or performance integrity for intralaminar matrix cracks, heat damage, delaminations (interlaminar matrix cracks) and fractured fibres (holes).

MATRIX CRACK ANALYSIS

As previously stated, the transverse or intralaminar matrix cracks have little effect on the strength of fibre dominated composite laminates but can reduce local stiffness of the laminate that can lead to changes in stress patterns of laminate. An assessment of the local loss of stiffness can be simply achieved by using a degraded value of the matrix stiffness and strength properties of each affected ply in the laminate stack. This degradation is generally set at about 60% to 80% of the material matrix dominated ply property value. The degree of degradation is applied directly to the ply matrix dominated properties or used in micromechanics analysis of the ply property. The laminate is then analysed with the typical structural analysis techniques with a change in the material ply database properties.

Recommended ply property degradation factors are provided in Table 4.2.

As an example of the degraded ply values listed in Table 4.2, Table 4.3 shows the effect on a unidirectional carbon fibre prepreg matrix dominated properties are reduced due to matrix cracking.

Poisson's ratio of the ply properties shows little change due to matrix degradation.

If the effect of stiffness reduction to the laminate is severe, particularly under compressive loads and fatigue cycling, then delaminations are likely to have already initiated. If matrix cracks are present with any of the other damage types, the influence of transverse matrix cracks is less significant, and therefore stress analysis of them is not required. The loss of structural integrity due to transverse matrix cracks associated with delamination becomes a secondary effect; however, the local change in stiffness due to the transverse matrix cracks should be taken into consideration when analysing the delamination criticality. Likewise, the transverse matrix cracks that might exist at the edge of a hole in a composite structure reduce the local stiffness and can reduce the effective stress concentration.

TABLE 4.2
Recommended Ply Property Degradation Factors (conservative)

Ply Pattern	Matrix Dominated Property Degradations Factor	Fibre Dominated Property Degradations Factor
Unidirectional Ply ($V_f > 40\%$)	80%	20%
Plain (Fine) Weave Fabric	30%	5%
Plain (Coarse) Weave Fabric	50%	10%
Twill Weave	40%	10%
4 Harness Satin Weave	40%	10%
8 Harness Satin Weave	60%	10%
Braided Weave	40%	10%
Fine Weave Modifying Factor (1)	0.85 × value above	0.95 × value above
Coarse Weave Modifying Factor (2)	1.20 × value above	1.05 × value above
Swirl Mat ($V_f < 30\%$)	80%	30%
Chopped Mat ($V_f < 20\%$)	90%	40%

Notes: 1. Fine weave has more strands or tows per inch (per mm). A fine weave would have say 50+ strands (tows) per inch or 2 strands per millimetre.
2. Coarse weave will have 12 or less strands per inch (½ strands per millimetre).

TABLE 4.3

Reduction of Ply Properties Due to Matrix Cracking

Property		Pristine Value	Reduction Factor	Degraded Value
Transverse Modulus	E_y (msi/GPa)	1.3/9.0	80%	0.26/1.8
In-Plane Shear Modulus	G_{xy} (msi/GPa)	1.0/7.1	80%	0.21/1.4
Transverse Tensile Strength	Y_T (ksi/MPa)	7.5/52	80%	1.5/10
Transverse Compression Strength	Y_C (ksi/MPa)	30/207	80%	6.0/41.4
In-Plane Shear Strength	S (ksi/MPa)	13.5/93	80%	2.7/19
Longitudinal Modulus	E_x (msi/GPa)	20/130	20%	16/104
Longitudinal Tensile Strength	X_T (ksi/MPa)	210/1,447	20%	168/1,158
Longitudinal Compression Strength	X_C (ksi/MPa)	210/1,447	20%	168/1,158

An example of the structural analysis of an intralaminar matrix cracked composite laminate is provided below and is based on the following information:

- A 24-ply CFRP laminate with the configuration of $[\pm45/0_2/90/0_2/\pm45/0_2/90]_s$, has the top 4 plies containing matrix cracks due to local heat damage. The damaged plies extend over a 500 sq. mm area.
- The laminate is fabricated from unidirectional plies with properties listed in Table 4.3.
- The laminated composite properties both in the pristine and degraded properties are provided in Table 4.4. Table 4.4 also shows the percentage loss of structural performance due to the damage.

DELAMINATIONS

The two fundamental analysis methods previously discussed, LEFM and point stress analysis, can be applied to delaminations in composite laminates. Again, however, the LEFM method requires evaluation of basic material property values relating to the two approaches considered. These material properties have been and currently are difficult to correlate to composite materials. This difficulty is due to several

TABLE 4.4

Composite Laminate Properties for a Matrix Damaged Structure

Property	Good	Cracked	% Difference
E_{o1} (GPa)	82.6	78.7	95%
E_{fl} (GPa)	74.5	67.1	90%
σ_{o1} (MPa) @–4,000 µstrain	330	315	95%
R_{FPF}	1.198	1.103	92%
R_{Ult}	2.555	2.409	94%

factors that make the resulting properties sensitive to large scatter. Factors such as ply stacking sequence, unidirectional tapes vs. cloth, fibre volume fraction, same ply orientation stacking, symmetry of the sub-laminate due to delaminations, etc., contribute greatly to the inconsistency of the material properties. Unfortunately, such material property data is difficult to come by and thus makes use of the LEFM a difficult and often an expensive undertaking.

The use of the point stress analysis method, which attempts to determine the stress state at the tip of the delamination crack front, is also at present difficult due to how the crack front opening displacement is analysed. The problem becomes more difficult when dealing with the mixed-mode stresses at the crack tip. However, it is worth noting that the in-plane stresses at the crack tip, when compared with their respective strength values, are usually insignificant after considering the out-of-plane stresses. It is these out-of-plane stresses that drive delamination crack growth, where the normal out-of-plane stress (σ_z) tends to dominate crack propagation.

We recall that it is the compressive strength which is severely reduced, and that tensile loading of delaminated composite laminates only reduces structural strength by typically less than 20%. The tensile strength reduction is due to local load share between the sub-laminates and an induced secondary bending moment. More on this outcome of the induced secondary bending will be discussed in a coming section of this chapter. It is under compressive loading that the crack has the potential to grow following sub-laminate buckling. For an initial estimate of the potential of delamination crack growth, we determine the stability of the sub-laminate under design loading conditions. Three methods are presented here for the analysis of delamination growth potential due to sub-laminate buckling.

Effective Laminate Stiffness

The effective laminate stiffness approach (O'Brien) determines the global change in laminate Young's modulus and defines the reduced buckling behaviour of an orthotropic homogeneous plate. By using a rule of mixtures approach, the effective laminate stiffness is calculated from

$$E_{\text{eff}_1} = \left(E^* - E_{\text{lam}} \right) \frac{A^*}{A} + E_{\text{lam}}$$

where

$$E^* = \frac{\sum_{i=1}^{m} \left(E_i t_i \right)}{t}, \text{ effective stiffness of the multiple delaminations}$$

m = the number of delaminations at a particular through-the-thickness point
E_i = the effective modulus of the ith sub-laminate
t_i = the thickness of the ith sub-laminate
t = total laminate thickness
A^* = area of delaminated region
A = total laminate panel surface area ($a \times b$)

E_{lam} = laminate axial Young's modulus = $\dfrac{1}{\left[A'_{11} \right] t}$, and

$$[A'] = [A]^{-1} + [A]^{-1}[B][D^*]^{-1}[B][A]^{-1}$$

$$[D^*] = [D] + [B][A]^{-1}[B]$$

$[A]$ = Laminate in-plane stiffness matrix (absolute)
$[D]$ = Laminate flexural stiffness matrix (absolute)
$[B]$ = Laminate coupling stiffness matrix (absolute)

The effective stiffness in the primary laminate direction is then used to determine the critical buckling stress, and this value is then compared to the applied design laminate stress at the global level.

Sub-Laminate Buckling Instability

In the sub-laminate buckling instability method, the individual sub-laminate loads are estimated by

$$N_{SL_{n_i}} = \frac{N_i E_{SL_{n_i}}}{\sum\limits_{n=1}^{m} E_{SL_{n_i}}}$$

where
      $i = 1,2,6$ (normal and in-plane shear directions)
      n = sub-laminate number
      m = number of sub-laminates
      N_i = applied in-plane axial and shear loads
      $E_{SL_{n_i}}$ = effective sub-laminate in-plane stiffnesses

The effective sub-laminate stiffness can be determined from

$$E_{SL_{n_i}} = \frac{1}{a_{ii}}$$

where

$$[a] = [a] + [A][B]\{[D] - [B][a][B]\}^{-1}[B][a], \quad \text{and}$$

$[a]$ = laminate in-plane compliance matrix
$[B]$ = laminate in-plane coupling stiffness matrix
$[D]$ = laminate in-plane flexural stiffness matrix

The critical buckling load for each sub-laminate assumes that the coupling flexural matrix has no effect on the buckled shape and that the buckled shape is represented by

$$w(x,y) = \frac{\left(a^2 - x^2 - y^2\right)^2}{a^4} d_o$$

where

d_o = the magnitude of the central deflection
a = the diameter of the delamination being assessed

So, then the critical sub-laminate buckling load is determined from

$$N_{crSL_n} = \frac{315}{17a^2} D_{ij}$$

where

$$D_{ij} = D_{11} + 1.29D_{12} + D_{22} + 0.889D_{66}$$

Hence, the sub-laminate load is compared to the laminate critical buckling load to estimate if buckling is likely to occur. Obviously, the larger the delaminated size gets (a^2), the more critical buckling load reduces.

Buckling of Laminates Plates

A simply supported laminated plate under uniform biaxial compression loading (Figure 4.11) has the following expression for the critical longitudinal axial buckling load (Whitney, 1987):

$$N_o^{cr} = \frac{\pi^2 \left[D_{11}m^4 + 2(D_{12} + 2D_{66})m^2n^2R^2 + D_{22}n^4R^4 \right]}{a^2 \left(m^2 + kn^2R^2 \right)}$$

where

$$N_1^{cr} = -N_o^{cr}$$

k = Biaxial loading parameter = $\dfrac{N_2}{N_1} \Rightarrow N_2^{cr} = kN_1^{cr}$

a = panel length in the 1-direction
b = panel width in the 2-direction
R = panel aspect ratio = a/b
m & n = integer values representing the half sinewave number for critical buckling mode (Figure 4.12).

D_{ij} = Flexural stiffness matrix components = $\begin{bmatrix} D_{11} & D_{12} & D_{16} \\ D_{21} & D_{22} & D_{26} \\ D_{61} & D_{62} & D_{66} \end{bmatrix}$

The analysis is based on special orthotropic laminates. This assumes that $A_{16} = A_{26} = 0$, $[B] = 0$ and $D_{16} = D_{26} = 0$. The extension shear stiffness coupling coefficients A_{16} and A_{26} are zero with balanced laminates, and extension/flexural stiffness coupling coefficients $[B] = 0$ with laminates possessing mid-plane symmetry. However, the flexural shear stiffness coupling coefficients D_{16} and D_{26} are non-zero terms in typical symmetric and balanced laminates. The magnitude of the flexural shear stiffness coupling coefficients is much less than the major flexural stiffness coefficients D_{11} and D_{22} and can thus be considered of negligible influence.

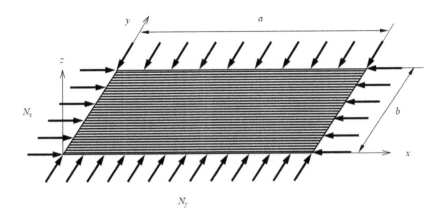

FIGURE 4.11   Rectangular laminated plate under in-plane biaxial compression loads.

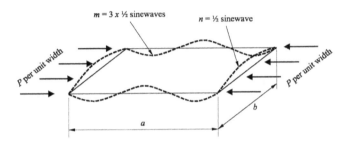

FIGURE 4.12   Panel buckling modes.

The application of this analysis to delaminations and disbonds is relatively straight forward. Consider a single-layer delamination of non-regular shape (Figure 4.13). A regular rectangular shape is superimposed on the region (Figure 4.13). This rectangular sub-laminate shape is analysed at the buckling panel. The boundary conditions of the buckling sub-laminate have been shown experimentally (Heslehurst, 1995) to be clamped at the loaded ends (*a*) and simply supported along the edges (*b*) (Figure 4.14). However, a simply supported boundary condition along all edges will be a conservative analysis.

A buckled sub-laminate is no longer orthotropic and is typically fully anisotropic (no symmetry or ply balance). However, the orthotropic solution is reached quickly with increasing ply numbers. This is illustrated in Figure 4.15. The reduced flexural stiffness matrix can be used to modify the critical buckling loads. The reduced flexural stiffness matrix considers the effects of an unbalanced laminate (A_{16} and A_{26} are non-zero) and mid-plane asymmetry $[B] \neq 0$.

The expression for the reduced flexural stiffness matrix is

$$[D^*] = [D] - [B][A]^{-1}[B]$$

The values of D_{ij} are substituted by the values of $[D^*]$ in the critical buckling expression.

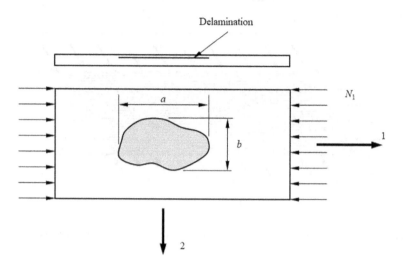

FIGURE 4.13   Delamination analysis geometry.

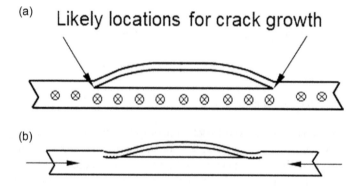

FIGURE 4.14   True delamination modelled cross-sectional shapes. (a) Transverse crack opening; (b) Axial crack opening.

HOLES AND FRACTURED FIBRES

The basic assumption of the fracture fibre stress analysis is that the chaotic fibre fracture condition (Figure 4.16) is machined into the clean hole. The hole shape is also assumed to be relatively simple, as a circular hole or an elliptical hole. In this chapter, three approaches to hole stress analysis are discussed: (1) in-plane analysis at the hole boundary; (2) estimation of the interlaminar stresses; and (3) a simple, handheld-calculator method of the hole stress state. All the methods assume that the hole is through-the-thickness. Partially through-the-thickness holes are more complicated as there is unsymmetrical laminate warping involved, by-pass loads and stress concentration complications to consider. This assumption will make the stress analysis of a through-the-thickness hole much simpler.

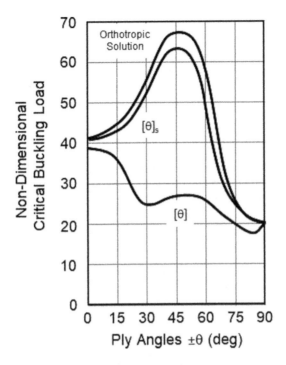

FIGURE 4.15 Buckling chart of anisotropic laminates. (Modified from Jones, R. M., *Mechanics of Composite Materials*, 2nd Edition, CRC Press, Boca Raton, FL, 1998.)

FIGURE 4.16 The jagged appearance of a fibre fractured hole (chaotic).

Hole In-Plane Stress Analysis

The in-plane stress analysis of holes requires a knowledge of the in-plane laminate stiffness matrix [A], hole geometry and applied far-field stress state. Seven methods of hole stress analysis are reported in the literature. In this book, we review four commonly applied methods based on the relative ease of application to holes in composite laminates. The methods are discussed in detail in the following paragraphs.

Average Stress Failure Criterion

The Average Stress Failure Criterion predicted the notch stress (σ_N) of the laminate at the hole edge. This is a stress that is averaged over a distance a_o from the hole edge, such that the ratio of notched to unnotched strength in a finite width laminate is given by

$$\frac{\sigma_o}{\sigma_N} = \frac{2(1-\rho)}{2-\rho^2-\rho^4-\left[K_T-3\right]\left[\rho^6-\rho^8\right]}$$

where

$$\rho = \frac{R}{R+a_o}$$

$a_o \approx 4.0$ mm (characteristic dimension) and is typical of high-performance composite materials

R = Hole radius

$$K_T = 1 + \sqrt{\frac{2}{A_{22}}\left[\sqrt{A_{11}A_{22}} - A_{12} + \frac{A_{11}A_{22}-A_{12}^2}{2A_{66}}\right]}$$

A_{ij} = in-plane laminate stiffness matric for i,j = 1, 2 and 6

For an orthotropic composite material, we can express K_T in terms of the laminate engineering constants

$$K_T = 1 + \sqrt{2\left[\sqrt{\frac{E_2}{E_1}} - \upsilon_{21}\right] + \frac{E_2}{G_{12}}}$$

where
  E_1 = Longitudinal Young's Modulus of the laminate
  E_2 = Transverse Young's Modulus of the laminate
  G_{12} = In-plane Shear Modulus of the laminate
  υ_{21} = Major Poisson's Ratio of the laminate

The predicted failure due to the presence of a hole in a composite laminate occurs over an average distance a_o from the hole edge. The ratio of notched to unnotched strength in a finite width laminate is given by:

Point Stress Failure Criterion

The Point Stress Failure Criterion is a modification of the Average Stress Failure Criterion discussed above. The failure stress due to the presence of a hole in a composite laminate is predicted at a fixed distance d_o from the hole boundary, such that the notched to unnotched stress ratio is

$$\frac{\sigma_o}{\sigma_N} = \frac{2}{2 + \rho^2 + 3\rho^4 - \left[K_T - 3\right]\left[5\rho^6 - 7\rho^8\right]}$$

where

$$\rho = \frac{R}{R + d_o}$$

$d_o \approx 1.0$ mm (characteristics dimension)
$K_T =$ as above

Figure 4.17 plots the stress distributions of both the average stress failure criterion and the point stress failure criterion from the hole edge. Note that average stress failure criterion and the point stress failure criterion only evaluate the stress concentration on a circular hole under an axial tension load. The stress concentration is determined at one point on the hole, perpendicular to the loading direction.

Greszczuk Method

The Greszczuk method provides the in-plane stress solution to a complex in-plane loading condition around the circumference of a circular hole in a composite laminate (Greszczuk). The failure initiation points can be determined anywhere on a

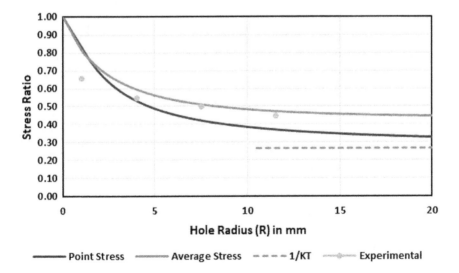

FIGURE 4.17   Stress distribution perpendicular to the hole.

circular hole boundary. The method is derived from the solution of the orthotropic plate governing differential equations in terms of the Airy Stress function and elastic compliance matrix. The Greszczuk method estimates the in-plane stress state around the boundary of a circular hole subjected to biaxial and in-plane shear stresses. The formulation of the Greszczuk stress concentration method is given through the following equations and in association with Figure 4.18.

The local stress around the hole boundary in the 1-2 plane is

$$\sigma_{1\theta} = \sigma_\theta \sin^2\theta, \quad \sigma_{2\theta} = \sigma_\theta \cos^2\theta \quad \tau_{12\theta} = -\sigma_\theta \sin\theta\cos\theta$$

where

θ = the angle from the x-axis, positive CCW

$$\sigma_\theta = \text{Re}\left[\frac{(\sigma_1+\sigma_2)\mu - \sigma_2\rho + \tau_{12}\lambda}{\left[1+\beta_1^2 - 2\beta_1\cos(2\theta)\right]\left[1+\beta_2^2 - 2\beta_2\cos(2\theta)\right]}\right],$$

the real term to the solution of a complex number result.
and

σ_1, σ_2 & τ_{12} = far-field in-plane stresses

$$\mu = \left[1+\beta_1\right]\left[1+\beta_2\right]\left[1+\beta_1+\beta_2 - \beta_1\beta_2 - 2\cos(2\theta)\right]$$

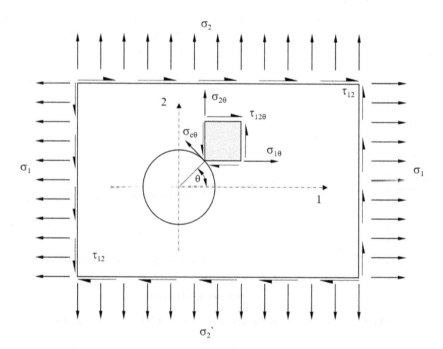

FIGURE 4.18   Greszczuk orthotropic laminate stress concentration method geometry.

$$\rho = 4\left[\beta_1 + \beta_2 - \{1 + \beta_1\beta_2\}\cos(2\theta)\right]$$

$$\lambda = 4\left[\beta_1\beta_2 - 1\right]\sin(2\theta)$$

$$\beta_1 = \frac{\alpha - 1}{\alpha + 1} \quad \beta_2 = \frac{\delta - 1}{\delta + 1}$$

$$\alpha = \sqrt{G + \sqrt{G^2 - E}} \quad \delta = \sqrt{G - \sqrt{G^2 - E}}$$

$$G = \frac{E_2}{2G_{12}} - \upsilon_{12} \quad E = \frac{E_2}{E_1}$$

and

E_1, E_2, G_{12} & υ_{21} are the elastic constants derived from the laminate in-plane stiffness matrix $[A_{ij}]$, such that

$$\upsilon_{21} = \frac{A_{12}}{A_{22}} \text{ major} \quad \upsilon_{12} = \frac{A_{12}}{A_{11}} \text{ minor}$$

$$E_1 = A_{11}\left(1 - \upsilon_{21}\upsilon_{12}\right) \quad E_2 = A_{22}\left(1 - \upsilon_{21}\upsilon_{12}\right)$$

$$G_{12} = A_{66}$$

Note that the Greszczuk method does not include a requirement to have a characteristic dimension or a step away from the hole edge to account for a free edge singularity problem.

Tan-Tsai Method

The Tan-Tsai method offers hole stress concentration analysis of composite laminates in a complex in-plane far-field stress state around an elliptical hole that can be orientated at any angle to the principal loading direction. The Tan-Tsai method is based on the complex potential analysis and the method of superposition of stresses. The following formulation is derived from the hole geometry in Figure 4.19. The stress field due to the opening is

$$\sigma_{n_1} = \text{Re}\left[\frac{\mu_1^2 f_1 g_2 - \mu_2^2 f_2 g_1}{\Delta\mu}\right] \quad \sigma_{n_2} = \text{Re}\left[\frac{f_1 g_2 - f_2 g_1}{\Delta\mu}\right]$$

$$\sigma_{n_6} = \text{Re}\left[\frac{\mu_2 f_2 g_1 - \mu_1 f_1 g_2}{\Delta\mu}\right] \quad \Delta\mu = \mu_1 - \mu_2$$

where

$$f_j = \frac{1 - i\mu_j\lambda}{\beta_j\sqrt{\beta_j^2 - 1 - \mu_j^2\lambda^2} + \beta_j^2 - 1 - \mu_j^2\lambda^2}$$

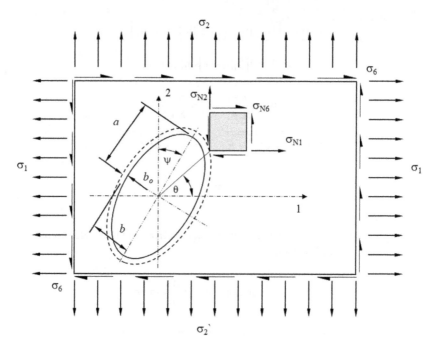

FIGURE 4.19 Tan-Tsai stress formulation geometry.

$$g_j = i\lambda\sigma_{o_1} - \mu_j\sigma_{o_2} + \left[i\mu_j\lambda - 1\right]\sigma_{o_6} \quad \text{for } j = 1,2$$

$$\beta_j = (1+\alpha)\cos\theta + \mu_j(\lambda+\alpha)\sin\theta$$

$$\lambda = \frac{b}{a} \quad \alpha = \frac{b_o}{a}$$

$2a$ = ellipse major axis $2b$ = ellipse minor axis

b_o = characteristic dimension
y = angle around the ellipse CCW from the +ve x-axis

$$\sigma_{o_1} = \sigma_1 \sin^2\Omega + \sigma_2 \cos^2\Omega + \sigma_6 \sin(2\Omega)$$

$$\sigma_{o_2} = \sigma_1 \cos^2\Omega + \sigma_2 \sin^2\Omega - \sigma_6 \sin(2\Omega)$$

$$\sigma_{o_6} = (\sigma_2 - \sigma_1)\cos\Omega\sin\Omega - \sigma_6 \cos(2\Omega)$$

where
Ω = angle of the ellipse major axis from the 2-axis

$$\sigma_I = \text{far-field stresses} \quad \text{for } i = 1,2,6$$

μ_j is the complex roots of the laminate characteristic equation, such that the characteristics equation is in terms of the laminate compliance matrix $[a_{ij}]$

$$a_{11}\mu^4 - 2a_{16}\mu^3 + \left[2a_{12} + a_{66}\right]\mu^2 - 2a_{26}\mu + a_{22} = 0$$

Generally, the majority of composite laminate are orthotropic, such that

$$a_{16} = 0 \text{ and, } a_{26} = 0$$

so that the reduced characteristic equation gives the solution of the roots as

$$\mu_j = \sqrt{\frac{\pm\sqrt{[2a_{12} + a_{66}]^2 - 4a_{11}a_{22}} - [2a_{12} + a_{66}]}{2a_{11}}}$$

or

$$\mu_j = \frac{i}{2}\left[\sqrt{\frac{E_1}{G_{12}} - 2\upsilon_{21} + 2\frac{E_1}{E_{12}}} \pm \sqrt{\frac{E_1}{G_{12}} - 2\upsilon_{21} - 2\frac{E_1}{E_2}}\right]$$

The final stress distribution due to the opening is then given by

$$[\sigma_N] = [\sigma_o] + \text{Re}[\sigma_n]$$

The circumferential stress on the periphery of the hole can also be calculated from the following stress transformation equation:

$$\sigma_\theta = \frac{\sigma_{N_1}C^2\sin^2\theta + \sigma_{N_2}D^2\cos^2\theta - 2\sigma_{N_6}CD\sin\theta\cos\theta}{\Gamma}$$

where

$$C = a + b_o$$

$$D = b + b_o$$

$$\Gamma = C^2\sin^2\theta + D^2\cos^2\theta$$

INCLUSION METHODOLOGY OF STRESS CONCENTRATION IN AN ORTHOTROPIC MATERIAL

An extension of the Tsai-Tan method is to have a solid material in the hole. This is referred to as an inclusion model of the stress concentration and is akin to a filled hole with, say, an unloaded fastener. The analysis used Figure 4.20 and the following analysis approach.

Using the compliance matrix data for the composite plate and the inclusion as

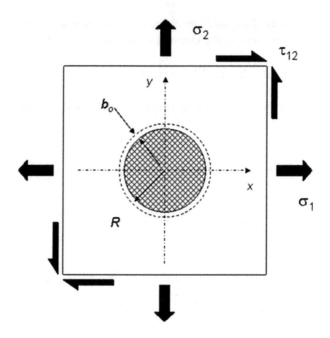

FIGURE 4.20   Inclusion model geometry.

$$a_{ij} = \begin{bmatrix} a_{11} & a_{12} & 0 \\ a_{21} & a_{22} & 0 \\ 0 & 0 & a_{66} \end{bmatrix} \quad \text{composite plate}$$

$$a'_{ij} = \begin{bmatrix} a'_{11} & a'_{12} & 0 \\ a'_{21} & a'_{22} & 0 \\ 0 & 0 & a'_{66} \end{bmatrix} \quad \text{inclusion}$$

From the characteristic equation of the composite laminate, we define the following parameters:

$$\kappa = -\mu_1\mu_2 = \sqrt{\frac{a_{22}}{a_{11}}}$$

$$v = -1(\mu_1 - \mu_2) = \sqrt{\frac{2a_{12} + a_{66}}{a_{11}} + 2\sqrt{\frac{a_{22}}{a_{11}}}}$$

The following composite laminate and inclusion interaction constants are defined from these two parameters above and the composite laminate and inclusion compliance matrix coefficients

$$F_1 = (a_{11}a_{22} + a'_{11}a'_{22})\kappa + a_{22}(a_{66} + 2a'_{12}) + (a_{11}a'_{22}\kappa + a_{22}a'_{11})v - (a_{12} - a'_{12})^2\kappa$$

$$F_2 = \frac{\left(a_{11}a_{22}+a_{11}'a_{22}\right)}{\kappa} + a_{11}\left(a_{66}+2a_{12}'\right)+\left(\frac{a_{22}a_{11}'}{\kappa}+a_{11}a_{22}'\right)\frac{\nu}{\kappa} - \frac{\left(a_{12}-a_{12}'\right)^2}{\kappa}$$

$$F_6 = a_{11}\kappa\nu + \left(a_{66}'+2a_{12}\right)\kappa + a_{22}\left(2+\nu\right)$$

Now the inclusion stress state is determined from

$$\sigma_1' = \frac{\sigma_{o_1}}{F_1}\left[a_{11}a_{22}\left(\kappa+\nu\right)+a_{11}a_{22}'\kappa\left(1+\nu\right)+a_{22}\left(a_{12}+a_{66}+a_{12}'\right)\right]$$

$$+\frac{\sigma_{o_2}}{\sigma_{o_2}}\left[a_{11}\left(a_{22}-a_{22}'\right)+a_{22}\left(a_{12}-a_{12}'\right)\frac{\left(\nu+\kappa\right)}{\kappa^2}\right]$$

$$\sigma_2' = \frac{\sigma_{o_2}}{F_2}\left[a_{11}a_{22}\frac{\left(1+\nu\right)}{\kappa}+a_{22}a_{11}'\frac{\left(\kappa+\nu\right)}{\kappa^2}+a_{11}\left(a_{12}+a_{66}+a_{12}'\right)\right]$$

$$+\frac{\sigma_{o_1}}{F_1}\left[a_{22}\left(a_{11}-a_{11}'\right)+a_{11}\left(a_{12}-a_{12}'\right)\kappa\left(1+\nu\right)\right]$$

$$\sigma_6' = \frac{\sigma_{o_6}}{F_6}\left[a_{11}\kappa\nu+\left(2a_{12}+a_{66}\right)\kappa+a_{22}\left(2+\nu\right)\right]$$

For a circular inclusion $\lambda = 1$

$$g_j = i\left(\sigma_{o_1}-\sigma_1'\right)-\mu_j\left(\sigma_{o_2}-\sigma_2'\right)+\left[i\mu_j-1\right]\left(\sigma_{o_6}-\sigma_6'\right) \quad \text{for } j=1,2$$

$$f_j = \frac{1-i\mu_j}{\beta_j\sqrt{\beta_j^2-1-\mu_j^2}+\beta_j^2-1-\mu_j^2} \quad \text{as before with } \lambda=1$$

$$\beta_j = (1+\alpha)\left[\cos\theta+\mu_j\sin\theta\right] \quad \text{as before with } \lambda=1$$

$$\alpha = \frac{b_o}{a}$$

a = hole radius b_o = characteristic dimension

θ = angle around the ellipse CCW from the +ve x-axis
The local inclusion notch stress is thus

$$\sigma_{n_1} = \text{Re}\left[\frac{\mu_1^2 f_1 g_2 - \mu_2^2 f_2 g_1}{\Delta\mu}\right] \quad \sigma_{n_2} = \text{Re}\left[\frac{f_1 g_2 - f_2 g_1}{\Delta\mu}\right]$$

$$\sigma_{n_6} = \text{Re}\left[\frac{\mu_2 f_2 g_1 - \mu_1 f_1 g_2}{\Delta\mu}\right] \quad \Delta\mu = \mu_1 - \mu_2$$

Stresses due to the uniform stress field are as before

$$\sigma_{o_1} = \sigma_1 \sin^2 \Omega + \sigma_2 \cos^2 \Omega + \sigma_6 \sin(2\Omega)$$

$$\sigma_{o_2} = \sigma_1 \cos^2 \Omega + \sigma_2 \sin^2 \Omega - \sigma_6 \sin(2\Omega)$$

$$\sigma_{o_6} = (\sigma_2 - \sigma_1) \cos\Omega \sin\Omega - \sigma_6 \cos(2\Omega)$$

where

Ω = angle of the hole axis from the 2-axis

$$\sigma_i = \text{far-field stresses} \quad \text{for } i = 1, 2, 6$$

The final stress distribution due to the opening is then given by

$$[\sigma_N] = [\sigma_o] + \text{Re}[\sigma_n]$$

Figure 4.21 illustrates the effect of a circular hole in a graphite/epoxy composite without an inclusion and with either an aluminium and steel inclusion. The effect of different composite materials with or without an inclusion is shown in Figures 4.22 and 4.23.

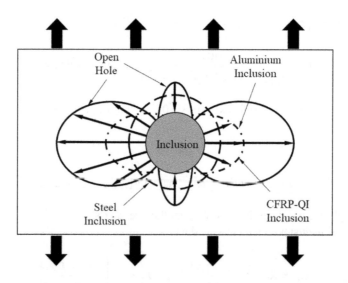

FIGURE 4.21 Circumferential stress distribution in a quasi-isotropic graphite/epoxy composite with and without an inclusion. (Redrawn from Tsai, S. W., *Composite Design*, 4th Edition, Thick Composites, Dayton, OH, 1988.)

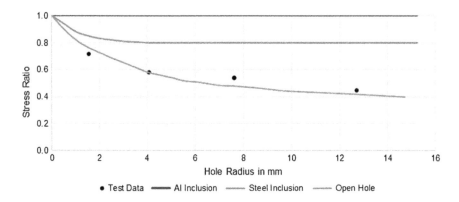

FIGURE 4.22 Strength reduction in a quasi-isotropic graphite/epoxy composite with and without an inclusion. (Modified from Tsai, S. W., *Composite Design,* 4th Edition, Thick Composites, Dayton, OH, 1988.)

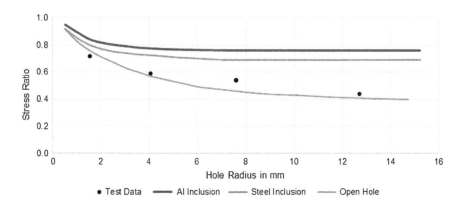

FIGURE 4.23 Strength reduction in a quasi-isotropic glass/epoxy composite with and without an inclusion. (Modified from Tsai, S. W., *Composite Design,* 4th Edition, Thick Composites, Dayton, OH, 1988.)

CONCLUSION

Stress analysis of the damaged area will provide information that will indicate the severity of the damage under a known loading condition(s). A composite or bonded component's structural integrity due to degradation must not be guessed or estimated but calculated. With a calculated value of the stress state and the significance of the damage with respect to the structural integrity, the engineer will be able to design the most appropriate repair scheme. The damage stress analysis will thus lead to either a cosmetic type, semi-structural (damage tolerance restoration) or a structural repair. Throughout the repair design, the restoration and maintenance of the local stiffness will be necessary to ensure appropriate load transfer through the repair patch. Hence, strength requirements and stiffness of the parent and repair patch become very important from this stage on in the repair of composite structures.

INTRODUCTION

5 Generic Repair Schemes

INTRODUCTION

The basic repair schemes for composite structures can be simplified to four types. Whilst each of the four types can have many variants, the fundamental description of each of the four types is the same. These four fundamental types of composite repair schemes are described in detail in this chapter.

OVERVIEW OF GENERIC REPAIR SCHEMES

The four basic levels of generic repair designs are listed as:

- Filling and sealing the damaged area as an environmental seal (termed as a cosmetic repair).
- Filling the damaged area and applying a doubler patch either as a bonded or riveted repair scheme (termed as a semi-structural repair).
- Bonding a flush patch to the damaged area as a scarf repair scheme (termed as a bonded structural repair).
- Bolting a patch to the damaged area as a doubler repair scheme (termed as a bolted structural repair).

Each of these generic repair schemes is explained in detail in the following paragraphs.

FILLING/SEALING REPAIR SCHEME

SURFACE DAMAGE REPAIR

When the structural significance of the damage is negligible, then the main requirement is to develop a repair for environmental protection. This type of repair scheme is simply a cosmetic repair. For such a repair scheme, the damage may not necessarily be removed, but moisture removal is still required. The damaged area is filled with a suitable potting compound (neat resin or mixed with chopped fibres) and then possibly sealed with a layer of fibreglass/epoxy woven cloth, Figure 5.1.

EDGE DAMAGE REPAIR

Specific edge damage to the sides of a composite laminate is also commonly seen as mild abrasion, low-level scratches and surface resin degradation. There are two basic types of repair to minor edge damage, which are either resin filling and/or low modulus edge strap, see Figure 5.2.

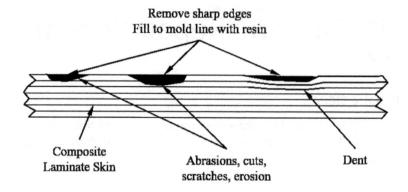

FIGURE 5.1 Surface damage cosmetic repair (non-structural).

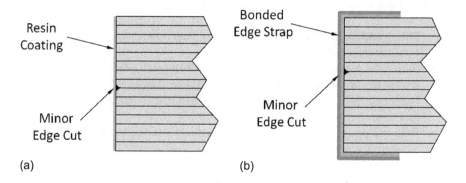

FIGURE 5.2 Edge damage minor repair (resin and/or strap).

Fastener Hole Damage Repair

Minor fastener hole damage such as bearing surface abrasion can be repaired with a low viscous resin coating and then reamed. A couple of examples are illustrated in Figures 5.3, 5.4 and 5.5.

FILLING/DOUBLER PATCH REPAIR SCHEME

Doubler Repair Scheme – Non-Structural Filler

As the structural severity of the damage increases, particularly for thin skins and honeycomb sandwich panels, some load transfer over the damaged region will be required. Such a filling/doubler patch repair scheme is both cosmetic and semi-structural, Figure 5.6. The damaged materials are usually removed, leaving a simple geometric hole (partial or through-the-thickness). Absorbed moisture must be removed when high-temperature repair fabrication is done so as not to cause poor repair scheme quality. The repair filling is generally of a low modulus material (so as not to attract unwanted loads), and the repair doubler patch is bonded to the parent

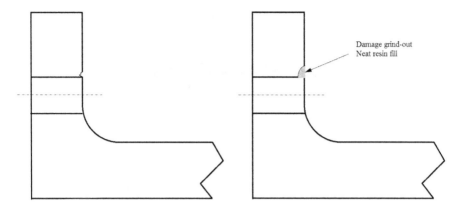

FIGURE 5.3   Neat resin sealing after surface grind out (for ≤0.005 inches deep scratches).

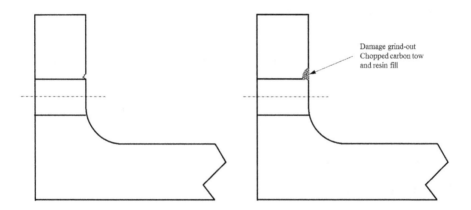

FIGURE 5.4   Resin and chopped carbon tow filling after surface grind out.

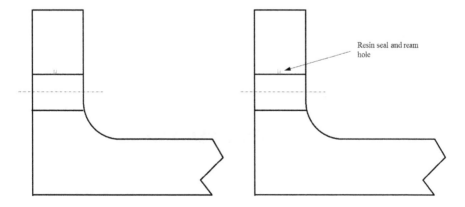

FIGURE 5.5   Neat resin sealing for delaminations ≤0.040 inches deep from hole edge.

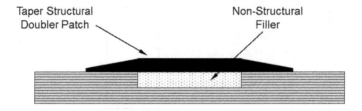

FIGURE 5.6   Semi-structural doubler repair scheme with a non-structural filler.

skin and filler material. As the load is transferred from the parent laminate into the doubler and out again into the parent laminate, the filler material should not draw much load from the doubler patch.

DOUBLER REPAIR SCHEME – STRUCTURAL FILLER

In order to reduce the protruding thickness of the structural doubler, a structural filler repair scheme can be used. In the structural filler-doubler repair scheme (Figure 5.7), the load is transferred from the parent structure into the doubler strap and then into the filler material, then back into the doubler strap and finally into the parent material. The structural filler is essentially a hole-filling material. The filler also reduces the stress concentration issues likely on the damage area of the parent material. Such repair schemes are typical for significant, partial through-the-thickness damage of a moderately thick laminate.

FLUSH BONDED PATCH REPAIR SCHEME

LAMINATE REPLACEMENT

In relatively thin structures that have been significantly reduced in strength by the presence of damage, a flush bonded patch repair scheme will provide the greatest strength restoration where aerodynamic smoothness is essential. The repair process is to remove the damage and carefully scarf or step the hole out; again, drying the laminate prior to repair is important. The patch, designed and cut to fit in the hole, can be either pre-cured and secondarily bonded or co-cured to the damaged area. Co-cured patches are generally stronger. A doubler patch is also included in the

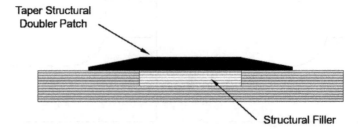

FIGURE 5.7   Semi-structural doubler repair scheme with structural filler.

repair scheme as a sealing patch for the flush patch. The doubler patch is no more than four plies in thickness and can be of a lower modulus material. Figure 5.8 illustrates typical flush bonded patch repairs.

BOLTED EXTERNAL DOUBLER

The bolted patch repair is restricted to thicker laminated sections that require ample structural integrity to be restored. However, the full structural strength is unlikely to be achieved, but restoration of the design load carrying capacity can be designed with a bolted patch repair. The bolted patch repair can be via a semi-flush or double-lap repair scheme, depending on the design requirements, and is typical of that shown in Figure 5.9. The repair process is to remove the damage and create a hole with circular ends, remove any moisture, drill the locating fastener holes in the parent laminate, and attach the inner-, flush- and outer-patch panels. The patch panels and fasteners should be coated with a sealing compound and fitted wet.

SANDWICH STRUCTURE REPAIR

There are six basic levels of generic repair scheme designs. There are:

- surface damage only (cosmetic) – filling and sealing the damaged area
- facing damage (semi-structural or structural repair) – facing replacement
- facing and core damage (cosmetic) – core replacement and cosmetic skin repair
- facing and core damage (semi-structural) – core replacement and applying a doubler repair patch
- facing and core damage (structural) – core replacement and bonding a flush repair patch
- facing and core damage (structural) – core replacement and bolted repair patch

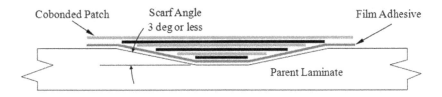

FIGURE 5.8 Flush bonded patch repairs.

FIGURE 5.9 Bolted patch repair.

Surface Damage Only (Cosmetic)

When the significance of the damage is small to the surface skin, then the main requirement for repair is the environmental protection of the underlying structure. This only then needs a cosmetic-type repair scheme. For the cosmetic repair scheme, the damage may not necessarily have to be removed. However, it is important to remove any absorbed moisture or other contaminating substance. Absorbed moisture could present long-term problems such as corrosion of metallic core or core long-term environmental degradation or repair scheme separation. The damaged area will need to be blended out for metallic skins, but for composite skins, surface damage is typically filled with a suitable potting compound (neat resin or mixed with chopped glass fibre) or just a resin overcoat. Bonding a simple glass fibre/resin woven cloth patch is also suggested for composite skins to provide a final seal of the repaired area. Figure 5.10 provides an illustrative example of the cosmetic surface repair scheme.

Facing (Skin) Damage (Semi-Structural or Structural Repair)

Where the skins require a degree of damage tolerance restitution, the application of a doubler patch is all that is required. The doubler patch is sized (dimensioned) based on the required overlap length to transfer the load from the parent structure to the repair patch. Also, the adhesive needs to be selected that provides the required load transfer strength. Figure 5.11 illustrates the semi-structural patch over the damaged area. Note that the damaged skin is removed, and the replacement plug can be either structural or non-structural. For a structural plug, the bond overlap length must be determined to ensure the adhesive is not overloaded.

If a structural repair is used to restore structural integrity of the skin, then a bonded flush repair is preferred for composite skins. Metal skins will typically use a bolted repair patch. The scarf angle for composite skin repairs is about 1 degree that is represented by a 0.25″ ply drop-off rate. The damage skin is machine-scarfed, and the repair plies are laid down with the smallest ply down first (Figure 5.12). A flush metal skin repair is typical for semi-structural repair requirements (Figure 5.13).

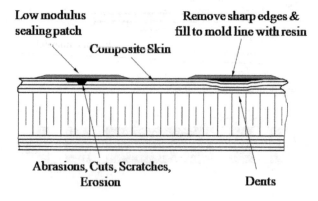

FIGURE 5.10   Cosmetic repair of skin damage (non-structural).

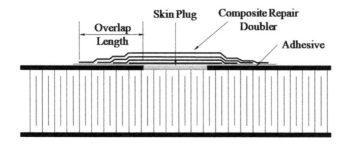

FIGURE 5.11   Semi-structural bonded doubler repair of composite skin damage.

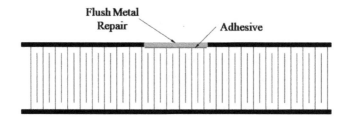

FIGURE 5.12   Structural bonded scarf repair of composite skin damage.

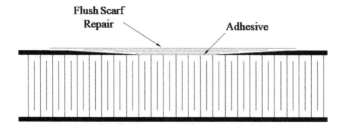

FIGURE 5.13   Semi-structural bonded flush repair of metal skin damage.

Surface preparation of the metal is essential for durability of the repair patch to the core. Typically, a single filler ply is included at the core interface. For bolted repair in metal skins, the bolt is sized on the skin thickness, and typical spacing of the bolts will be an edge distance of two bolt diameters and spacing between adjacent bolts will be three bolt diameters or more (see Figure 5.14).

FACING (SKIN) AND CORE DAMAGE (COSMETIC)

Cosmetic repairs to both skin and core follow similar requirements to the skin-only cosmetic repair but will include core replacement or core filling. Such a repair scheme is depicted in Figure 5.15. Of particular interest is to ensure that the core is bonded effectively to the undamaged inner skin face and spliced to the original core. Good engineering practice is to replace the core with the same or slightly higher density and for honeycomb core to match the ribbon direction.

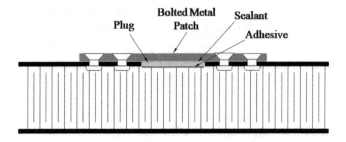

FIGURE 5.14   Structural bolted doubler repair of metal skin damage.

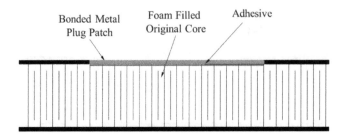

FIGURE 5.15   Cosmetic skin and core repair.

FACING (SKIN) AND CORE DAMAGE (SEMI-STRUCTURAL)

As the structural severity of the damage increases, particularly for thin skins and honeycomb sandwich panels, some load transfer over the damaged region will be required to restore damage tolerance integrity. Such a repair scheme (Figure 5.16) is both cosmetic and semi-structural. The damage is usually removed and so is any absorbed moisture. Damaged honeycomb core is then replaced or a foaming resin plug is inserted. Over the repair region, a doubler patch is bonded in place. The hole plug should be of low modulus so as to not attract load; hence, the load is transferred from the parent laminate into the doubler and out again into the parent laminate. The

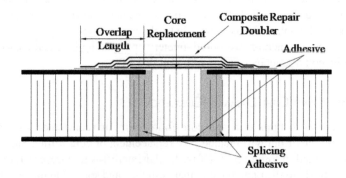

FIGURE 5.16   Semi-structural bonded doubler repair of composite skin damage with core replacement.

core should be replaced with a similar density or slightly higher density material. Note that the core must be effectively bonded to the undamaged skin (lower skin shown in the figure) so as not to result in a debond in the future. The core is also spliced with foaming adhesive to the parent core material. With honeycomb, it is good engineering practice to align the replacement core ribbon direction with that of the original core.

Facing (Skin) and Core Damage-Structural Repair

In relatively thin structures, where damage has significantly reduced the strength, a flush bonded patch repair scheme will provide the greatest strength restoration. A flush bonded repair scheme, as shown in Figure 5.17, will also provide aerodynamic or operational smooth contours. The repair process is to remove the damage and carefully scarf or step the hole out. Such repairs are typical of composite skins. Since most repairs will be with high temperature curing adhesive, drying the laminate prior to applying the repair scheme is essential. As a general rule, the scarf angle is achieved by scarfing with a taper of 0.25″ (6 mm) per ply. Scarf repairs are generally a co-bonded composite repair patch. The core is bonded to the lower undamaged skin of Figure 5.17 but ensure that the core is effectively bonded to the undamaged skin. The core replacement is spliced to the parent core with the honeycomb core ribbon direction aligned.

Including the core replacement, a bolted patch repair is restricted to thicker skins or metallic skins that require ample structural integrity to be restored. However, the full structural strength is unlikely to be achieved due to the influence of the fastener load carrying capabilities. However, restoration of the design load carrying capacity can be provided with a bolted patch repair. The bolted patch repair can be via a semi-flush or double-lap repair scheme, depending on the design requirements, and is typical of that shown in Figure 5.16 and with core replacement of Figure 5.18. The repair process is to remove the damage and create a hole with circular ends, remove any moisture, drill the locating fastener holes in the parent laminate, and attach the inner-, flush- and outer-patch panels. The patch panels and fasteners should be coated with a sealing compound and fitted wet. The outer repair scheme material is typically a metal as this provides a better material to drill the holes and clamp the repair patch. A stiffer repair patch can be used to reduce patch thickness.

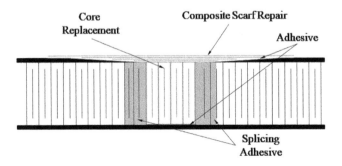

FIGURE 5.17   Flush bonded patch repairs.

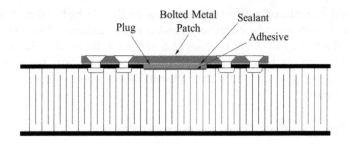

FIGURE 5.18   Bolted external patch repairs.

6 Composite Joining Methods and Design Requirements

INTRODUCTION

Unlike the joining of metal structures, the requirements for geometric sizing of a composite, mechanically fastened joint are often more critical, particularly when the joint is bearing primary loads. This concern stems from the relative weak shear and bearing strengths of composite laminates. For adhesively bonded joint design, the free ends of the joint are the critical locations where particular care must be taken to ensure peeling stresses are kept low and the traction-surface has the correct ply orientation. This chapter provides a detailed understanding of the unique requirements for the mechanical fastening of composite structures and how to design adhesively bonded joints to allow full load transfer, which will be related to the repair of composite structures.

The question most often asked is whether to use mechanical fasteners or to use adhesively bonded composite structural repair joint. This chapter provides design and analysis details that will clearly answer this question. Understanding the design and stress analysis of composite structural joints will allow participants to develop an understanding of preliminary design sizing and joint efficiency of composite repair joints.

A breakdown of the various joining methods is provided in Figure 6.1, where the basic joining methods are classified as mechanical fasteners, adhesively bonded and welded.

The most commonly used joining methods in composite repair are bolted or bonded repair schemes. Each repair scheme is basically one of the following (see Figure 6.2):

- Doubler joint
- Unsupported single-lap joint
- Unsupported single-strap joint
- Tapered unsupported single-strap joint
- Double-lap joint
- Double-strap joint
- Tapered double-strap joint
- Scarf joint (single and double)
- Stepped-lap joint (single and double)

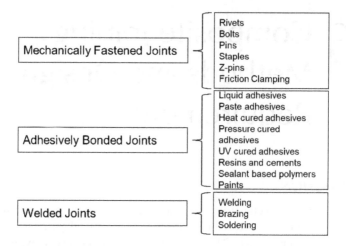

FIGURE 6.1 Types of structural joints.

BONDED VS. BOLTED JOINTS

ADVANTAGES AND DISADVANTAGES

In joining processes and configurations for composite materials, there is a number of advantages and disadvantages. These advantages and disadvantages are listed in Table 6.1.

MATERIAL PROPERTIES

In joint design and for the repair of composite structures, the material properties of both the parent structure and the repair patch, plus the adhesive and fasteners, are required.

Parent Structure and Repair Patch

An example of the properties for some typical metals used in repair and for two composite laminates in a quasi-isotropic configuration is given in Table 6.2. Typical high performance adhesive properties are shown in Table 6.3, and a few composite qualified fastener properties are listed in Table 6.4.

Shear Modulus:

$$G = \frac{E}{2(1+\upsilon)} \quad \text{only for isotropic materials}$$

Adhesive Properties

The adhesive properties of the materials in Table 6.3 are based on the idealised shear stress/strain curve (Figure 6.3). The idealised shear stress/strain curve represents the same shear strain energy as the actual shear stress/strain curve in Figure 6.3.

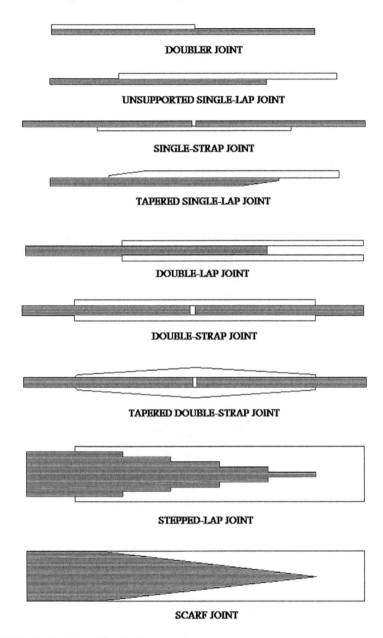

DOUBLER JOINT

UNSUPPORTED SINGLE-LAP JOINT

SINGLE-STRAP JOINT

TAPERED SINGLE-LAP JOINT

DOUBLE-LAP JOINT

DOUBLE-STRAP JOINT

TAPERED DOUBLE-STRAP JOINT

STEPPED-LAP JOINT

SCARF JOINT

FIGURE 6.2   Bolted and bonded joint configurations.

Fastener Properties

The fundamental properties for fasteners are the shear strength and tensile strength. The cross-sections of two composite dedicated fasteners are shown in Figures 6.4 and 6.5. The shear strength and tensile strength of Huck blind bolts and Cherry blind bolts are provided in Tables 6.4 and 6.5.

TABLE 6.1
List of Advantages and Disadvantages in Composite Joint Types

Joint Type	Advantages	Disadvantages
MECHANICAL FASTENERS (Bolts, Rivets, Screws, Pins, Staples, etc.)	Straight forward design Inspectable Repairable Any thickness Can be disassembled	Many parts Stress concentration Relative weaker joint Fatigue prone Must seal the joint Prone to fretting Prone to corrosion
ADHESIVE BONDING	Few parts Full load transfer Repairable Fatigue resilient Sealing Stiff connection Light-weight structure Smooth contour Corrosion resistant No stress concentrations	Difficult to inspect Surface preparation Environmental effects New design methods Trade skills required Thickness limited Residual stresses Cannot be disassembled Shear loading only

TABLE 6.2
Typical Adherend and Repair Patch Properties

Material Property	Aluminium 2025-T3	Titanium 6Al-4V	E-Glass/ Epoxy (QI)	Graphite/ Epoxy (QI)
Elastic Modulus (GPa)	72.4	**110.3**	19	54.8
Tensile Yield or FPF Strength (MPa)	331	1,000	72.9	320
Tensile Ultimate Strength (MPa)	448	1,103	412	587
Poisson's Ratio	0.33	0.31	0.27	0.28
CTE ($\times 10^{-6}/°C$)	22.5	8.82	11.3	1.87
Shear Strength (MPa)	276	690	60.8	227
Bearing Strength @ e = 2D (MPa)	903	1,872	72.9	320
T-T-T/Interlaminar Tensile Strength (MPa)	441	1,103	31	52
T-T-T/Interlaminar Compression Strength (MPa)	317	1,117	118	206
T-T-T/Interlaminar Tensile Modulus (GPa)	72.4	110	8.3	9.0

FPF – First Ply Failure.
T-T-T – Through-the-Thickness.

TABLE 6.3
Typical Structural Adhesive Properties (Standard Cure, Dry Condition, Room Temperature)

Material Property	FM 300	FM 73	EA 9628	EA 9696
Maximum Shear Strength (MPa)	34.5	48.3	77.9	31.9
Thickness (mm)	0.127	0.127	0.127	0.127
Elastic Strain	0.09	0.10	0.132	0.10
Plastic Strain	0.33	0.45	0.587	0.495
Effective Tensile Modulus (GPa)	7.34	5.96	2.38	1.92
Effective Tensile Strength (MPa)	93.1	86.2	51.7	47.2
Cure Temperature (deg C)	177	121	121	121

TABLE 6.4
Huck Blind Bolts (100° Flush, 130° Flush/Shear and Protruding)

Diameter in (mm)	All Titanium		CRES A-286 Sleeve/Titanium Pin	
	Shear Strength (kN)	Tensile Strength (kN)	Shear Strength (kN)	Tensile Strength (kN)
1/8 (3.18)	5.0	2.7		
5/32 (3.97)	8.1	4.0	8.38	5.1
1/4 (6.35)	13.0	6.2	13.0	7.5
3/8 (9.53)	22.3	9.3	22.3	12.9
5/16 (7.94)	32.1	13.8	32.1	18.5

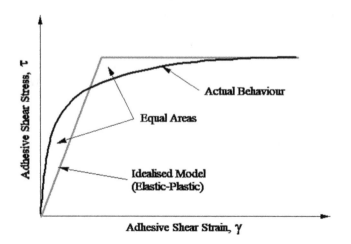

FIGURE 6.3 Idealised adhesive shear stress/strain curve.

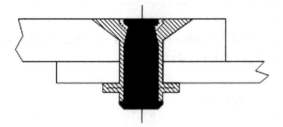

FIGURE 6.4   Huck Ti-Matic blind bolt 100° flush head.

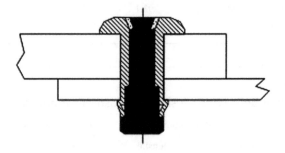

FIGURE 6.5   Cherry Maxibolt blind bolt protruding head.

TABLE 6.5
Cherry Blind Bolt (Maxibolt) (100° Flush and Protruding)

	Alloy Steel		CRES A-286	
Diameter in (mm)	Shear Strength (kN)	Tensile Strength (kN)	Shear Strength (kN)	Tensile Strength (kN)
1/8 (3.18)			5.4	3.0
5/32 (3.97)	10.4	6.0	8.8	5.1
1/4 (6.35)	15.3	9.3	13.0	7.5
3/8 (9.53)	26.2	16.2	22.2	12.9
5/16 (7.94)	37.8	23.1	32.0	18.5

COMPOSITE JOINT DESIGN HINTS

The following recommendations should be used for good joint design in composite materials:

- For bearing strength limitations, 2–3D edge and 3–4D pitch distances between bolts are required.
- Bonded joints of thick sections should use a stepped-lap joint rather than a scarf joint, as the former provides more consistent results and better design flexibility.

- Never have 90° plies in contact with the other adherend in a bonded joint. This will reduce lap shear strength significantly. Note that the effect of 8-harness satin warp/fill face down on the bondline will have a lesser reduction in strength.
- To reduce the peak stresses in a bonded joint use ±45° plies at the faying surface.
- Design the bonded joint stronger than the adjacent structure.
- Bonding works best for thin structures.
- Surface preparation of a bonded joint is most critical.
- Have the adhesive lightly stressed to reduce creep.
- Size bonded overlaps to the hot/wet environment.
- Taper the ends of a bonded joint to reduce peel stresses.
- Adhesives work best in shear and are poor in peel, but composites are weaker in interlaminar tension than the peel strength of the adhesive.

ADHESIVELY BONDED JOINTS

The analysis of adhesively bonded structural joints is achieved from an understanding of bonded joint mechanics. The types of joints, the joint geometry and material properties are all that is required to conduct the analysis. The shear stress in all the joint types must be considered, and for the single and double-lap joint configurations, the out-of-plane peel stresses should also be estimated.

ADHESIVE CHARACTERISTICS

The adhesive properties provided later are based on the idealised shear stress/strain curve (Figure 6.3), which gives the adhesive an elastic/plastic behaviour. Three parameters that characterise the adhesive shear properties are (see Figure 6.6):

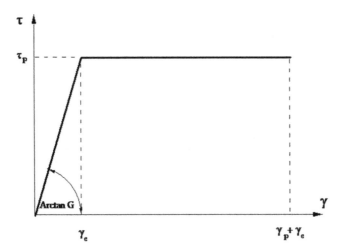

FIGURE 6.6   Idealised adhesive shear stress/strain curve.

- the plastic shear stress (maximum shear stress) τ_p
- the elastic shear strain γ_e
- the plastic shear strain γ_p

The shear stiffness G in the linear region is derived from two of these parameters

$$G = \tan^{-1}\left(\frac{\tau_p}{\gamma_e}\right)$$

The structural properties of any adhesive are temperature dependent. As the temperature increases, the adhesive becomes more viscoelastic, with corresponding changes in the mechanical properties, in particular a marked drop in the maximum shear stress, a significant increase in the plastic shear strain and a slight decrease in the shear stiffness (Figure 6.7). However, the percentage change in strain energy is not as dramatic, of the order of 10% to 20%. The effect can be taken into account through the factor of safety applied to the overlap length. Likewise, there is a decrease in the viscoelastic behaviour of the adhesive, with a drop in operational temperature. Table 6.6 provides a typical example of adhesive property changes with temperature.

During the design of an adhesively bonded repair scheme, the temperature range must be determined and the corresponding adhesive shear and peel performance used to determine the joint overlap length geometry and adhesive load carrying capacity.

ADHESIVELY BONDED JOINT FAILURE MODES

The six common failure modes in adhesively bonded joints can be classified as failure in the adherends (far-field and composite interlaminar), cohesive (shear and peel) and adhesive (shear and peel). Each of the six failure modes is illustrated in Figures 6.8 through 6.13. Through good surface preparation processes, the adhesive failure modes (bondline interfacial fracture) can be eliminated. With good joint design analysis, the joint load carrying capacity will be greater than the failure field strength; this means that the joint structural efficiency will be greater than 100%.

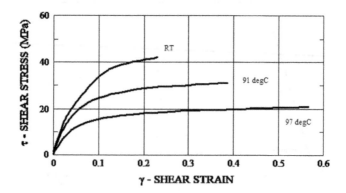

FIGURE 6.7   The effect of temperature on the shear stress/strain curve of FM 300.

TABLE 6.6
Adhesive Mechanical Property Variations with Operational Temperatures

Property	150 deg F	RT	0 deg F
Maximum Shear Strength τ_p (MPa)	25.9	34.5	44.8
Initial Shear Modulus G_{init} (MPa)	647	1,014	1,400
Maximum Shear Strain γ_{max}	0.45	0.3	0.12
Elastic Shear Strain γ_e	0.056	0.055	0.045
Plastic Shear Strain $\gamma_p = \gamma_{max} - \gamma_e$	0.394	0.245	0.075
Peel Young's Modulus E_c (GPa)	5.58	31.0	36.2
Peel Strain ε_c	0.0161	0.0107	0.0043
Peel Strength σ_c (MPa)	89.6	110	155
Adhesive Ductility Parameter γ_p/γ_e	7.0	4.45	1.7
Shear Strain Energy $\tau_p(\gamma_e/2 + \gamma_p)$	1.58 (+16%)	1.36	0.63 (−54%)

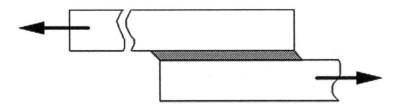

FIGURE 6.8 Adherend failure (tension).

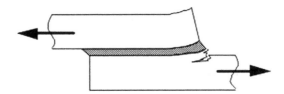

FIGURE 6.9 Adherend failure (interlaminar).

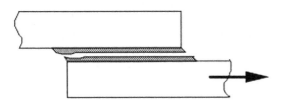

FIGURE 6.10 Cohesive shear failure.

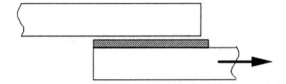

FIGURE 6.11   Adhesive shear failure.

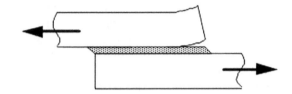

FIGURE 6.12   Adhesive peeling failure.

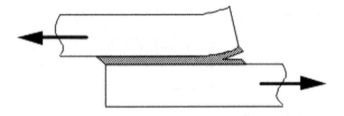

FIGURE 6.13   Cohesive peeling failure.

The remaining three failure modes will be analysed for bondline shear loads (cohesive shear) and bondline peel stresses (cohesive peel and composite interlaminar fracture).

ADHESIVELY BONDED DOUBLE-LAP JOINT

The stress state of the adhesively bonded double-lap joint examines the shear strength capacity of the adhesive in the joint and the induced peeling stresses. The following simply provides the design equations and design charts.

The shear load per unit width in the adhesive and the overlap length of a double-lap joint can be determined from Figure 6.14.

In a thermally matched and stiffness-balanced adhesively bonded double-lap joint, the adhesive allowable shear load per unit width is given by

$$P_{\text{adhesive-shear}} = 2\tau_{\text{av}}l$$

$$P_{\text{adhesive-shear}} = 4\sqrt{E_o t_o}\sqrt{\tau_p h_a\left(\frac{\gamma_e}{2} + \gamma_p\right)}$$

where

l = overlap length
$E_o t_o$ = outer adherend effective stiffness
h_a = adhesive thickness

with

$\alpha_i = \alpha_o$, thermally matched
$E_i t_i = 2E_o t_o$, stiffness balanced
 i – inner adherend (or for repair design p … parent structure)
 o – outer adherend (or for repair design r … repair patch)

Along with the effects of adherend stiffness imbalance and thermal expansion mismatch, Figures 6.15 and 6.16 provide the graphical analysis solution.

With stiffness imbalance and thermal mismatch and with a sufficiently long overlap length (typically $l > 25t_o$), the adhesive shear load carrying capacity is the lesser

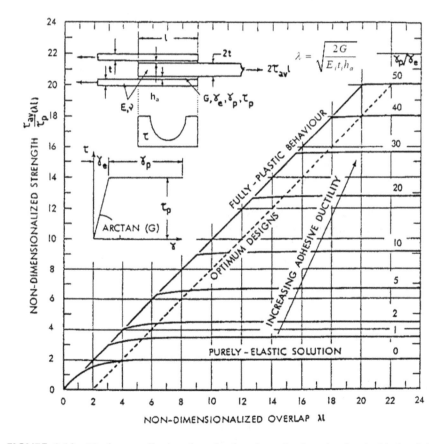

FIGURE 6.14 Maximum adhesive shear load and overlap length of a double-lap joint. (Modified from Hart-Smith, L.J, Adhesive Bonded Double-Lap Joints, NASA CR-112235, 1973.)

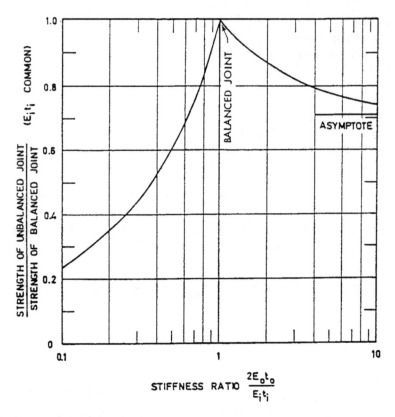

FIGURE 6.15   Strength reduction of double-lap joints due to adherend stiffness imbalance. (Courtesy of Hart-Smith, L.J, Adhesive Bonded Double-Lap Joints, NASA CR-112235, 1973.)

of the following design expressions (each expression represents the load state at the end of the double lap joint outer adherends):

$$P_{\text{adhesive-shear}} = \frac{4\tau_p}{\lambda}\left[\sqrt{1+2\frac{\gamma_p}{\gamma_e}+\text{CTHERM}(1)}\right]\left[\frac{1+\text{ETR}(1)}{2}\right], \quad \text{or}$$

$$P_{\text{adhesive-shear}} = \frac{4\tau_p}{\lambda}\left[\sqrt{1+2\frac{\gamma_p}{\gamma_e}+\text{CTHERM}(2)}\right]\left[\frac{1+\text{ETR}(2)}{2}\right]$$

where $\text{ETR}(1) = \dfrac{E_i t_i}{2 E_o t_o}$, and $\text{ETR}(2) = \dfrac{2 E_o t_o}{E_i t_i}$

$$\text{CTHERM}(k) = \frac{(-1)^{k+1}(\alpha_o - \alpha_i)\Delta T \lambda}{\tau_p\left[\dfrac{2}{E_i t_i}+\dfrac{1}{E_o t_o}\right]}, \quad k = 1,2$$

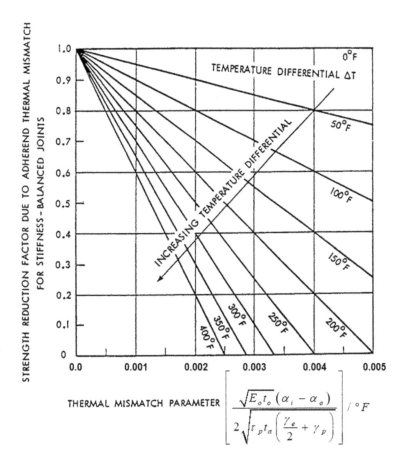

FIGURE 6.16 Strength reduction of double-lap joints due to adherend thermal mismatch. (Modified from Hart-Smith, L.J, Adhesive Bonded Double-Lap Joints, NASA CR-112235, 1973.)

$$\lambda = \sqrt{\frac{G}{h_a}\left[\frac{2}{E_i t_i} + \frac{1}{E_o t_o}\right]}$$

Although a double-lap bonded joint is usually symmetric, there are peel stresses induced due to local load path eccentricity. Figure 6.17 provides a design chart for estimating the induced peel stress in a symmetric double-lap adhesively bonded joint. The peel stress in a double-lap joint can be determined numerically through the following expression, which is based on the effective through-the-thickness modulus of the bonded joint:

$$\sigma_{c_{max}} = \tau_p \sqrt[4]{\frac{3E_c'\left(1 - \upsilon_{21_o}\upsilon_{12_o}\right)t_o}{E_o h_a}}$$

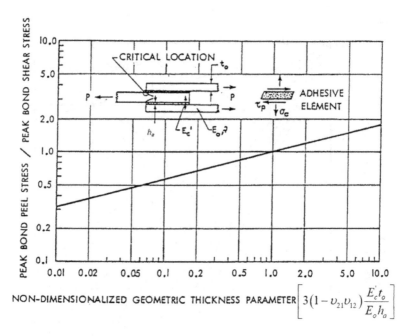

FIGURE 6.17 Peel stresses in double-lap bonded joints. (Modified from Hart-Smith, L.J, Adhesive Bonded Double-Lap Joints, NASA CR-112235.)

where

$$E_c' = \left[\frac{1}{E_c} + \frac{4}{E_i} + \frac{2}{E_o} \right]^{-1}$$

ADHESIVELY BONDED SINGLE-LAP JOINT

In addition to the two failure modes of primary interested discussed above for double-lap adhesively bonded joints, i.e. adhesive shear and peeling stresses, the single-lap joint introduces a third failure mode. Due to the load eccentricity generated by external load misalignment, a secondary bending stress is introduced into the adherends. The adherend failure mode is more common with thin adherends and that typically found in external patch repairs. This section will provide the design outcome for this third failure mode of interest in single-lap adhesively bonded joints.

The effective applied tensile load per unit width (P) of a single-lap joint that is stiffness-balanced ($E_1 t_1 = E_2 t_2$) and thermally matched ($\alpha_1 = \alpha_2$) is determined from Figure 6.18. The total maximum adherend stress is a combination of the far-field axial load and the induced secondary bending stress:

$$\sigma_{max} = \sigma_{axial} + \sigma_{bending\,max}$$

$$\sigma_{max} = \left[\frac{P}{t_1} + \frac{6M_1}{t_1^2} \right] = \left[\frac{P}{t_2} + \frac{6M_2}{t_2^2} \right] = \sigma_{av} \left[1 + 3k \left(1 + \frac{h_a}{t_1} \right) \right]$$

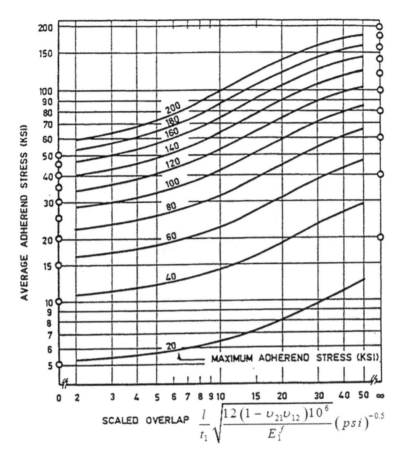

FIGURE 6.18 Load carrying capacity of a single-lap bonded joint due to secondary bending. (Modified from Hart-Smith, L.J, Adhesive Bonded Double-Lap Joints, NASA CR-112236, 1973.)

where

1 – is the parent structure for a repair scheme design

2 – the repair patch

$$k = \left[1 + \frac{\xi l}{2} + \frac{(\xi l)^2}{24}\right]^{-1}$$

$$\chi = \xi l$$

with

$$\xi^2 = \frac{P}{D}$$

$$D_1 = \frac{E_1 t_1^3}{12\left(1 - \upsilon_1^2\right)}, \quad \text{isotropic materials}$$

$$D_{11} = \frac{E_1^f t_1^3}{12\left(1 - \upsilon_{21}\upsilon_{12}\right)}, \quad \text{orthotropic materials}$$

ADHESIVELY BONDED SCARF JOINT

In order to maintain the high structural efficiencies in thick adherends, scarf and stepped-lap joints offer the only acceptable adhesively bonded joint solution. Using appropriate shallow scarf angles, the shear stress distribution in the adhesive is uniform for identical adherends ($E_1 t_1 = E_2 t_2$) and ($\alpha_1 = \alpha_2$). A simple stress analysis approach in such cases of identical adherends is shown using Figure 6.19.

The adhesive shear stress (τ) that is parallel to the scarf and the out-of-plane or normal stress (σ) that is perpendicular to the scarf is given by

$$\tau = \frac{P}{2t}\sin\left(2\theta\right)$$

$$\sigma = \frac{P}{t}\sin^2\left(\theta\right)$$

where
θ = the scarf angle
$l = t/\tan\theta$

As the scarf angle approaches zero, and by letting the joint shear stress $\tau = \tau_p$ of the adhesive, the normal stress approaches zero ($\sigma \longrightarrow 0$) and the adhesive allowable load is thus

$$P = Et\varepsilon_{\text{all}} = \frac{2\tau_p t}{\sin\left(2\theta\right)}$$

$$\text{therefore:} \quad \theta = \frac{\tau_p}{E\varepsilon_{\text{all}}}, \quad \text{in radians}$$

For commonly used structural adhesives, this requires a scarf angle of less than 3°. Generally, in composite repair, a 1° scarf angle is used. This represents a ply drop per 6 mm (0.25 in) for unidirectional plies.

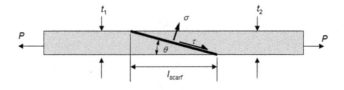

FIGURE 6.19 Simple scarf joint analysis.

When the adherends are not identical such that $(E_1 t_1 \neq E_2 t_2)$ and/or $(\alpha_1 \neq \alpha_2)$, then the effects of such can be expressed in the following expressions up to the elastic limit of the adhesive (γ_e):

$$P_{\text{Adhesive-Shear}} = \tau_p l \left\{ \text{ETR}(1) + \frac{[1 + \text{ETR}(1)]\text{CTHERM}(1)}{(\lambda l)} \right\}, \quad \text{left-side of the joint}$$

$$P_{\text{Adhesive-Shear}} = \tau_p l \left\{ \text{ETR}(2) + \frac{[1 + \text{ETR}(2)]\text{CTHERM}(2)}{(\lambda l)} \right\}, \quad \text{right-side of the joint}$$

where $\text{ETR}(1) = \dfrac{E_1 t_1}{E_2 t_2}$, and $\text{ETR}(2) = \dfrac{E_2 t_2}{E_1 t_1}$

$$\text{CTHERM}(k) = \frac{(-1)^{k+1}(\alpha_2 - \alpha_1)\Delta T \lambda}{\tau_p \left[\dfrac{1}{E_1 t_1} + \dfrac{1}{E_2 t_2} \right]}, \quad k = 1, 2$$

$$\lambda = \sqrt{\frac{G}{h_a}\left[\frac{1}{E_1 t_1} + \frac{1}{E_2 t_2} \right]}$$

The lesser value of $P_{\text{Adhesive-Shear}}$ is then taken.

The applied load per unit width of a scarf joint up to the elastic strain limit of the adhesive is shown in Figures 6.20 and 6.21 for both adherend stiffness imbalance and thermal expansion mismatch, respectively.

Apart from adhesive shear stress estimation, another limiting failure mode can occur in shallow angled scarf joints. The thin tip of the adherend scarf is prone to premature failure if the scarf angle is too small. This minimum scarf angle can be determined from:

$$\theta_{\min} > \frac{\tau_p}{F_u} \quad \text{(in radians)}$$

where F_u is the ultimate adherend strength (see Table 6.7).

This small angle means that the adherend tip will break off prior to the shear failure in the adhesive. This problem is more typical with long overlaps and/or brittle-type adhesives (a Ductile Parameter <4). The thin tip fracture problem can be overcome by using more ductile adhesives (a Ductile Parameter >9), effectively reducing the average shear stress across the bondline.

ADHESIVELY BONDED STEPPED-LAP JOINT

For a detailed analysis of a stepped-lap joint, a computer program is required because of the complexity of the analysis. However, for stepped-lap joints with greater than three steps, the scarf joint analysis is reasonably accurate. We must be aware that the last step in the stepped-lap joint is the critical region, so a 25% reduction in the load per unit width would certainly guarantee an adequate safety margin.

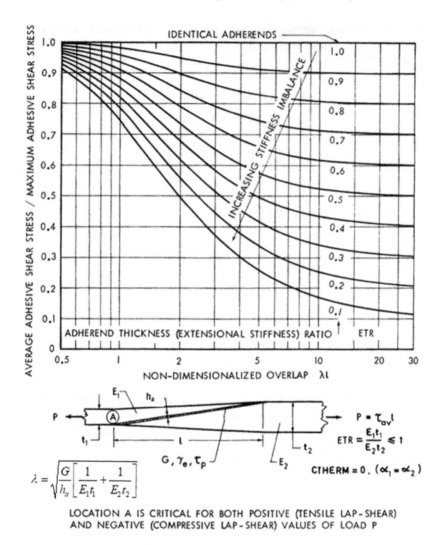

FIGURE 6.20 Scarf joint stiffness imbalance elastic strength. (Modified from Hart-Smith, L. J., Adhesive Bonded Scarf and Stepped-Lap Joints, NASA CR-112237, 1973.)

The use of stepped-lap joints has been used with some degree of success. From a design analysis point-of-view, these joints provide a more accurate stress assessment than a scarf joint, but they are rather difficult to fabricate, particularly with hand sanding. Recent developments with laser-guided ablative machining have been very successful, but the limitations of cost, location and equipment set-up remain a barrier to their use in composite repair workshops.

Stepped-lap joints have an advantage over uniform lap joints by having the adhesive strain concentration at the end of each step (Figure 6.22). When compared to scarf joints, the stepped-lap joints are easier to align and have far less adhesive thickness intolerances.

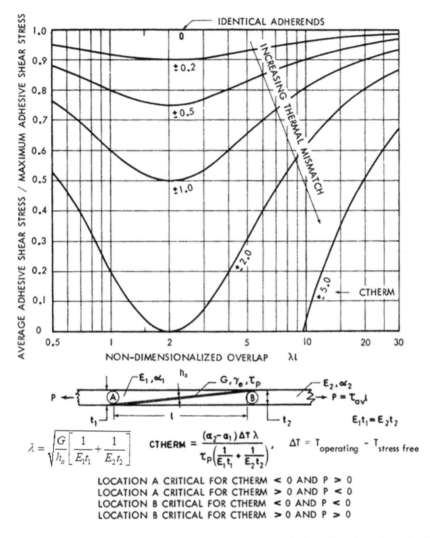

FIGURE 6.21   Scarf joint thermal mismatch elastic strength. (Modified from Hart-Smith, L. J., Adhesive Bonded Scarf and Stepped-Lap Joints, NASA CR-112237, 1973.)

Step-length optimisation for a stepped-lap joint is based on an idealised scarf angle (Figure 6.23). The angle of effective slope is given by

$$\theta_{\min} = \frac{\tau_p}{F_{ult}}$$

For preliminary design, an estimate of the total length of the stepped-lap joint is given by

$$l = \frac{P}{\tau_p}\left[\frac{E_2 t_2}{E_1 t_1}\right]$$

TABLE 6.7
Minimum Scarf Angles Against Thin Tip Fracture

Adherend	σ_{ult} (MPa)	Ductile Adhesive (τ_p = 41 MPa)	Brittle Adhesive (τ_p = 62 MPa)
Al 2024-T3	400	5.9°	8.8°
Al 7075-T6	552	4.3°	6.4°
Ti 6Al-4V	1103	2.1°	3.2°
QI G/E Composite	710	3.3°	5.0°
Stiff Composite	1379	1.7°	2.6°

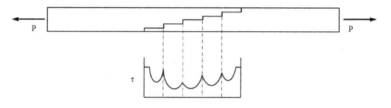

FIGURE 6.22   Stepped-lap joint and typical shear stress distribution.

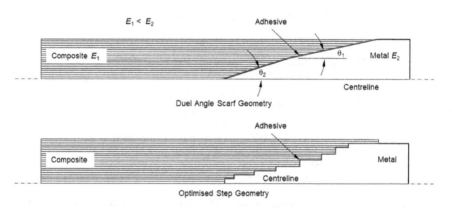

FIGURE 6.23   Taper angle optimisation of the effective scarf angle for stepped-lap joints. (Redrawn from Hart-Smith, L. J., Adhesive Bonded Scarf and Stepped-Lap Joints, NASA-CR-112237, 1973.)

OTHER ADHESIVELY BONDED JOINT CONSIDERATIONS

Bonded Joint Design Criteria Requirements

The design requirements for adhesively bonded joints are summarised in the following points:

- Bonded joints shall outlast the life of metal parts.
- Surface preparation must be reliable.

- Failure of adhesive by creep rupture shall not be permitted.
- Employ an elastic trough in the shear stress distribution.
- Avoid uniformly strained joints (i.e. balanced scarf joints).
- Avoid or alleviate discontinuities in structure.
- Ensure that static loads do not exceed adhesive plastic strain capacity.
- Restrict adhesives to elastic strain limits for fatigue load concerns.
- Minimise peel stresses – take bond loads in shear.
- Joint details with generous manufacturing tolerances.
- Joint details appropriate to local load intensity.
- Minimise stiffness imbalance and thermal mismatch across the joint.
- Cutouts, joints and eccentricities are the most critical structural design details.
- Bonded joint analysis is non-linear, but the use of an elastic/plastic model for the adhesive is simple and effective.
- Durability of bonded joints arises from a lightly loaded elastic trough and concentrated load transfer at the ends.
- Complex joints require non-linear computer programs.

MECHANICALLY FASTENED JOINT DESIGN

MECHANICALLY FASTENED JOINT FAILURE MODES

The failure modes of mechanically fastened joints occur either in the adherends or with the fastener itself. The failure modes of mechanically fastened joints are illustrated in Figure 6.24. Failure of the fastener is based on the shear strength and shear area of the fastener; details on the shear strength of fasteners can be found in the appropriate fastener design handbook. Whereas the failure modes of the adherend, net-tension, shear and bearing are based on joint geometry parameters, adherend material properties and the effects of stress concentration.

The actual failure mode is generally not one type in particular but more often is a combination of several of the failure modes or a mixed-mode failure. The design allowable stresses of a mechanically fastened joint will ultimately depend on:

- Joint geometry (hole size, plate width, edge distance and thickness)
- Clamping area and pressure
- Ply fibre orientation
- The stress concentration, which is a function of ply orientation and hole geometry, as depicted in Figure 6.25
- Moisture and temperature conditioning
- The nature of the applied stress (static or cyclic, axial or complex, in-plane or out-of-plane)

As a general design rule, complex mechanically fastened joints can be simply modelled as single-lap or double-lap joints with a single row or multiple rows of fasteners.

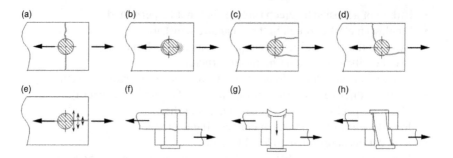

FIGURE 6.24 Failure modes of mechanically fastened joints. (a) Net-tension failure; (b) Bearing failure; (c) Shear out failure; (d) Combined shear and tension failure; (e) Cleavage failure; (f) Fastener shear failure; (g) Fastener pull through failure; (h) Fastener bending failure.

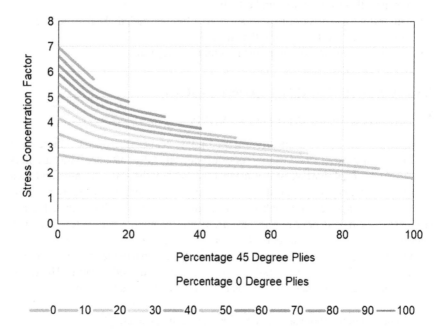

FIGURE 6.25 Open hole stress concentration factor in graphite/epoxy laminates.

SINGLE FASTENER ANALYSIS

Stress analysis of mechanically fastened joints with a single fastener is based on the geometry in Figure 6.26.

FASTENER SHEAR FAILURE

The shear stress on a fastener is given by

$$\tau_{\text{Shear}} = \frac{P_{\text{Allowable}}}{nA_{\text{Shear}}}$$

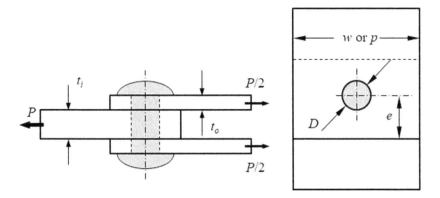

FIGURE 6.26 Single fastener geometry.

where

$P_{\text{Allowable}}$ = the allowable axial force

A_{Shear} = the fastener area subject to the shear force

$$= \frac{\pi D^2}{4}$$

D = fastener diameter

n = the number of shear planes, i.e. $n = 2$ for double shear.

Net-Tension Failure

The failure stress of the adherend due to the reduced cross-sectional area, net-tension, for metals is

$$\sigma_{\text{NT}} = \frac{P_{\text{Allowable}}}{\text{FoS}(w - D)t}$$

where

FoS = Factor of Safety

= 3 for static loading (typical)

= 4 for fatigue loading (typical)

w = width of specimen or fastener seam spacing

t = thickness of component under net-tension failure

There is a stress concentration effect, but due to local yielding, the effect is very localised and compensated by the Factor of Safety value. However, for a design consideration in composite materials, the stress concentration factor for a loaded hole k_{tc}, as a function of bolt diameter-to-panel width ratio, and edge distance-to-panel width ratio are necessary due to fluctuation in the stress concentration (see Figure 6.25):

$$\sigma_{\text{NT}} = \frac{k_{tc}P_{\text{Allowable}}}{(w - D)t}$$

where k_{tc}, at the hole 90° to the applied load is effectively

The general relationship between composite net-tension stress concentration factor k_{tc} and the elastic isotropic net-tension stress concentration factor k_{te} is given by

$$k_{tc} - 1 = C(k_{te} - 1)$$

where C is a constant of proportionality.

The approximate correlation of the constant of proportionality between composite net-tension stress concentration factor and the elastic isotropic net-tension stress concentration factor is that C is directly proportional to the percentage of 0-degree plies ($P0$) (see Figure 6.27):

$$C = \frac{P0}{100}$$

$$k_{te} = 2 + (\alpha - 1) - \frac{3}{2}\left[\frac{\alpha - 1}{\alpha + 1}\right]\Theta$$

where

$\theta = 1.5 - \dfrac{1}{2\beta}$ for $\beta \le 1$

$\theta = 1$ for $\beta > 1$

$\alpha = \dfrac{w}{D}$ and $\beta = \dfrac{e}{w}$

Shear-Out Failures

The stress analysis of shear-out failures in metallic structures is given by

$$\tau_{SO} = \frac{P_{\text{Allowable}}}{2et}$$

This relationship holds for composite laminates as long as the percentage of ±45° plies is significantly large. The optimum value is around 50% of the ±45° plies in the laminate.

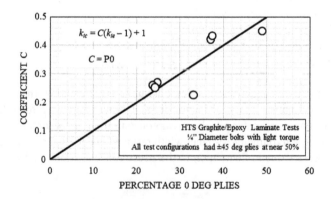

FIGURE 6.27   Stress concentration relief at fastener holes in composite laminates. (Redrawn from Hart-Smith, L. J., Bolted Joints in Graphite-Epoxy Composites, NASA-CR-144899, 1976.)

Bearing Failures

The stress due to bearing in composite structures is given by

$$\sigma_{BR} = \frac{k_{bc}P_{\text{Allowable}}}{Dt}$$

where $k_{bc} = \dfrac{k_{tc}}{(\alpha - 1)}$

Factors that affect bearing strength in composite laminates include:

- Clamping force of the torqued bolt
- Fastener fit on the hole
- Fastener bending stiffness
- Ply orientations
- Ply stacking sequence
- Environment effects (temperature and moisture)
- Countersink fastener angle
- Use of bushings
- Single- vs. double-lap joint configuration

MULTIPLE FASTENER ANALYSIS

To determine the most accurate fastener stress distribution in a multiple fastened panel typically a non-linear computer program is necessary. However, for estimates of mechanically fastened joint strength, the following can be assumed:

- The panel can be divided into strips of equal width having each fastener centrally located (Figure 6.28), where:

$$\text{Strip width } w = \text{Fastener pitch } p$$

$$\text{Strip load } P = \frac{P_T}{m} \text{ or, } P = \frac{P_T}{L}w$$

  where m = number of strips or fasteners in a row
- Each fastener in the strip will take a portion of the load (P), such that $P = \Sigma P_i$, where $i = 1$ to m. As a first estimate, let each fastener take an equal portion; therefore, fastener load is effectively

$$P_{\text{fastener}} = \frac{P_T}{\text{no. of fasteners}}$$

In reality, the first row of fasteners will take a greater proportion of the load due to joint flexibility, and each subsequent fastener row will take a reducing proportion. This is evidenced by the typical failure pattern of the first row of fasteners that is seen with bolted repair patches in composites, as shown in Figure 6.29.

The distribution of fastener loads in multi-rowed mechanically fastened joints is not linear, i.e. the first and the last rows of fasteners take a greater proportion of the

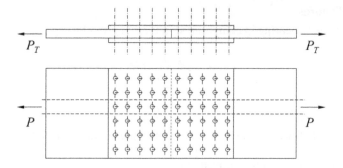

FIGURE 6.28   Multiple fastened joint – strip width approach.

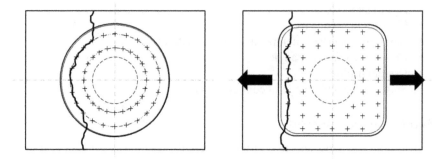

FIGURE 6.29   Failure of bolted repairs to a 24-ply graphite/epoxy laminate.

load than the next inboard row(s). This non-uniform load proportioning is due to the plate stiffness, the effective fastener bending stiffness and the ratio of the two plate stiffnesses being joined. With stiffer fasteners, relative to the plate stiffness, the load sharing becomes increasingly disproportionate.

As a preliminary method for determining fastener load distribution, the ratio of the fastener spring stiffness (K_f) to plate stiffness ($K = AE/l$) is initially determined. The effective fastener bending stiffness (K_f) or (spring stiffness) is estimated from

$$K_f = \frac{1}{C_f}$$

where
C_f = fastener effective compliance.

There are several models developed, and the following is the Boeing relationship:

$$C_f = \frac{1}{K_f} = \frac{2^{\left(t_i/D\right)^{0.85}}}{t_i}\left(\frac{1}{E_i} + \frac{3}{8E_f}\right) + \frac{2^{\left(t_o/D\right)^{0.85}}}{t_o}\left(\frac{1}{E_o} + \frac{3}{8E_f}\right)$$

Ef = fastener modulus of elasticity
D = fastener diameter

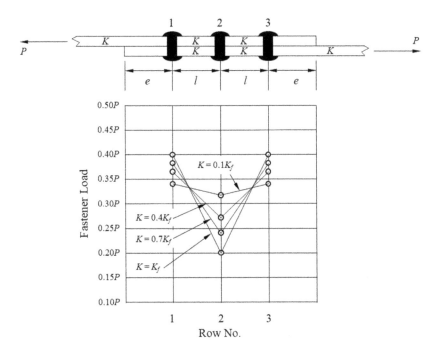

FIGURE 6.30 Fastener load distribution for three rows of fasteners (redrawn from Niu).

As an example of three-fastener rows, the fastener load share for an elastically balanced bolted joint is given by the following expressions, which are illustrated by Figure 6.30:

$$\text{RLF}_{R1} = \text{RLF}_{R3} = -0.0167 \left(\frac{K_f}{K_p} \right)^2 + 0.0803 \left(\frac{K_f}{K_p} \right) + 0.3362$$

$$\text{RLF}_{R2} = 0.0333 \left(\frac{K_f}{K_p} \right)^2 - 0.1607 \left(\frac{K_f}{K_p} \right) + 0.3275$$

OTHER CONSIDERATIONS FOR MECHANICALLY FASTENED JOINTS IN COMPOSITES

Other considerations that must be taken into account for the mechanical fastener designs in composite materials and structures are either process- or materials-orientated. These other design and fabrication considerations are as follows:

- In-plane or out-of-plane local stiffening of the hole area will improve the damage tolerance and increase the ultimate strength, but there are additional processing costs.
- The drilling of fastener holes in composites must be done with care so that back surface delaminations do not occur. Therefore, both the drill rotational

speed and feed speed must be controlled. Generally, we use a high drill rotational speed and a slow drill feed speed.
- In composite material, the use of interference fit fasteners should be avoided, as the interference fasteners will generate internal delaminations around the hole circumference. Fastener/hole tolerances should be of the order of 0.05 mm–0.1 mm (0.002″–0.004″).
- Countersunk fasteners in composite structures are a concern because there is the potential of delamination initiation. Typically, bushings or low-profile fastener heads are used rather than countersunk holes.
- The hole diameter-to-edge distance and fastener pitch distances recommended for composites structures are

$$\text{Edge distance } e = 2D \text{ to } 3D$$

$$\text{Pitch distance } p = 3D \text{ to } 4D$$

- Reduced clamping pressure is required but should not exceed the recommended value of $\sigma_z = 20$ MPa (2.9 ksi). Thus, the torque (including washer) is restricted to:

$$T = \frac{\sigma_z D^3}{1.658}$$

where
   a washer diameter $D_w = 2.2D$ is highly recommended
   a bolt torque coefficient constant of 0.2 is assumed
- Corrosion protection of the fastener is essential, particularly with the joining of dissimilar materials (i.e. carbon fibre composite to light-alloy metals).
- The composite laminate loaded hole strength is generally less than an unloaded hole. The load hole stress state is complex due to the bearing/by-pass load interaction. A conservative estimate for the bearing allowable strength in carbon/epoxy composites is 345 MPa (50 ksi) when test data is unavailable.

SUMMARY OF MECHANICALLY FASTENED JOINT ANALYSIS IN COMPOSITES

The general conclusions for mechanically fastened joints in composite materials are stated in the following points:

- The design of mechanically fastened joints can be relatively straight forward with simple joint configurations.
- Analysis of mechanically fastened joint strength requires substantial correction factors for several of the typical failure modes.

- For multi-rowed fastener joints, the bearing/by-pass load interactions need to be taken into account.
- The maximum joint strength is achieved by keeping the bearing stresses low, which can be done by closely spaced fasteners and multiple rows of fasteners.
- Optimum mechanically fastened joint geometries can be identified by the analysis discussed herein.
- Thick composites should be designed with low strain limits so that mechanically fastened joint repair is permitted, as illustrated in Figure 6.31.
- Mechanically fastened joint strength in composites can be improved with local reinforcement of the composite fastener hole.
- The best composite fibre orientation patterns are around (25% 0°, 50% ±45°, 25% 90°) to (37.5% 0°, 50% ±45°, 12.5% 90°) (see Figure 6.32).
- The linear elastic fracture behaviour of composite structures limits the static joint strength, particularly for mechanically fastened joints.

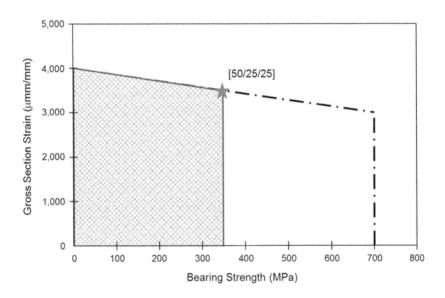

FIGURE 6.31 Graphite/epoxy allowable design limits to accommodate bearing stresses.

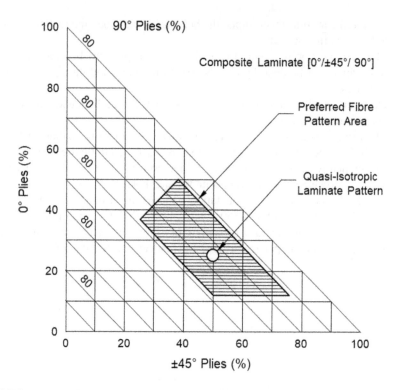

FIGURE 6.32 Preferred composite fibre patterns for mechanically fastened joints. (Redrawn from Hart-Smith, L. J., Adhesive Bonded Scarf and Stepped-Lap Joints, NASA-CR-144899, 1976.)

7 Repair Scheme Design

INTRODUCTION

SELECTION OF THE REPAIR METHOD

The repair design is primarily driven by non-engineering requirements such as:

- The repair facility capability
- The type of damage found
- Whether or not the repair scheme is to be installed on or off the aircraft
- The accessibility of the damaged area

Repair Facility Capability

The capability of suitable repair facilities has the strongest influence on the repair design. Without the appropriate tools, equipment and materials at hand, even the best repair designs cannot be installed. The level of repair a facility is authorised to undertake is dependent on the capability of the appropriate facilities. For example, flight line repairs are generally restricted to simple plug/patch repair types, whereas a depot-level repair should cover all repair types.

Types of Damage

The types of damage that have already been reviewed in terms of their structural significance often dictate the repair design. The damage types discussed in terms of repair design will be:

- Matrix cracks
- Delaminations
- Debonds
- Holes

On-Aircraft or Off-Aircraft

When repairing a damaged structural component, the question arises: should the component be removed. Some components, such as access panels and doors, can be easily removed and replaced by a serviceable component. The damaged door can then be repaired in the repair facility where appropriate environmental conditioning can take place. However, external skins on wings, fuselages, etc. cannot be removed and repairs must be done on the structure. There are two criteria which lead to repairs on the aircraft or to removing the component for repair; they are based upon:

- A comparison of time to remove and install the component to the time to repair it
- The hours to repair directly on the aircraft as opposed to removal, repair and re-installation of the component

111

Damaged Component Accessibility

The location of the damaged component is a serious limitation to repair design. If two-sided access is available, the repair design is often more effective. There are specific application methods which allow repairs to damaged components when one-sided access is only available. These will be discussed later.

REPAIR CRITERIA

The basis of the repair design follows logical repair criteria. The parameters of the repair criteria are listed in Table 7.1 and are detailed in the following paragraphs.

TABLE 7.1

Repair criteria

1. STATIC STRENGTH AND STABILITY
   Full versus Partial Strength Restoration
   Stability Requirements
2. REPAIR DURABILITY
   Fatigue Loading
   Corrosion
   Environmental Degradation
3. STIFFNESS REQUIREMENTS
   Deflection Limitations
   Flutter and Other Aeroelasticity Effects
   Load Path Variations
4. AERODYNAMIC/HYDRODYNAMIC SMOOTHNESS
   Manufacturing Techniques
   Performance Degradation
5. WEIGHT AND BALANCE
   Size of the Repair
   Mass Balance Effect
6. OPERATIONAL TEMPERATURE
   Low- and High-Temperature Requirements
   Temperature Effects
7. ENVIRONMENTAL EFFECTS
   Types of Exposure
   Effects to Epoxy Resins
8. RELATED ON-BOARD AIRCRAFT SYSTEMS
   Fuel System Sealing
   Lightning Protection
   Mechanical System Operation
9. COSTS AND SCHEDULING
   Downtime
   Facilities, Equipment and Materials
   Personnel Skill Levels
   Materials Handling
10. LOW OBSERVABLES

Static Strength and Stability

Any repair must be capable of supporting the design loads that are applied to the original structure. The two major aspects of this are:

- **Strength Restoration**. The first question to ask is if full strength restoration is required. The answer to this question is determined based on the results of the damage analysis.
- **Stability Requirements**. The greatest concerns in many of the damaged structures are instability under compressive loading and how to restore structural stiffness. The damage analysis will indicate where structural instability exists and the methods of designing a repair to overcome this instability.

Repair Durability

Any repair designed to restore the aircraft to flying conditions is generally expected to remain an integral part of the airframe for the aircraft's remaining service life (exceptions are rapid action type repairs). For commercial aircraft, their serviceable life is 50,000 flight hours, and for military aircraft, it is 4,000–6,000 flight hours, plus any life-of-type extensions. Thus, durability of the repair scheme must consider the following in its design phase:

- Fatigue loading of the structure and the effects on bolted and bonded joints, damage growth and monitoring the repair for continuing airworthiness assessment.
- Corrosion of components where dissimilar materials have been used in the repair and to ensure that corrosion protection precautions are still in place.
- Environmental degradation of resin type repairs is to be allowed for in the design of the repair, particularly moisture absorption and the hot/wet environment.

Stiffness Requirements

In aircraft structures where light-weight structures are an essential design requirement, stiffness is often more critical than strength. The same goes for repairs such that they must maintain the integrity of structural stiffness. The following must be considered in a stiffness repair design requirement:

- Deflection limitations of flying surfaces such as wings and flight controls are based on aerodynamic performances of the aircraft; repair should not unduly alter the aircraft's flying characteristics.
- Flutter and other aeroelasticity effects restrict the design of a repair so that its stiffness should be almost equal to the parent structure. Increased stiffness will decrease the flutter speed, and a decrease in stiffness can also change the flying characteristics.
- Load path variations are obviously undesirable in that areas within the structure will be loaded in excess of their design allowables. As a rule, the repair area stiffness should match that of the parent structure.

Aerodynamic/Hydrodynamic Smoothness

Aerodynamic/hydrodynamic smoothness is an important consideration when maximum speed or fuel efficiency is required. Those parts of the vehicle that require good aerodynamic/hydrodynamic smoothness, i.e. leading edges and where the boundary layer is laminar, must have flush or very thin external patch repair schemes. These repair types are based on local capabilities in manufacturing techniques, the effects of performance degradation, repair size and the possible effects of multiple damage sites.

Weight and Balance

The size of the repair and the local changes in weight can be insignificant to the total component weight, but in weight sensitive structures, such as control surfaces in dynamic flow fields, the effect to the mass balance can be highly significant. The effective change in local weight must be controlled to within certain limits, and in some cases, re-balancing of the component may be necessary.

Operational Temperature

The operating temperature influences the selection of repair materials, particularly adhesives and composite resins. Materials that develop adequate strength within the required operational temperature range must be selected. The combination of extreme temperatures with environmental exposure, the hot/wet condition, is often the critical condition for which the repair must be designed.

Environmental Effects

Composite and adhesive bonds are prone to significant degradation when exposed to various environments, such as fluids and thermal cycling. However, it is the absorbed moisture that is frequently the major long-term concern in terms of reduced durability of the repair design that must also be considered.

Related On-Board Vehicle Systems

The repair design must also be compatible with other on-board vehicle systems. Typically, these systems are as follows:

- **Fuel System Sealing**. In modern vehicles, the fuel is carried within the structure, i.e. in an aircraft wing and known as a "wet wing". Hence, any repair to wing skins which are in direct contact with the fuel system must seal the fuel tank, cater for out-of-plane fuel pressure forces and not contaminate the fuel system during the repair process.
- **Lightning Protection**. If electrical conductivity in the parent structure has been required for lightning protection, then the repair must also incorporate the same degree of electrical conductivity.
- **Mechanical System Operation**. Any component that is required to move during the operation of the vehicle or is in close proximity to a moving component, and is subsequently repaired must ensure that the repair does not impede component operation. For example, aircraft retracting flaps must be repaired such that the repair still provides the adequate retraction clearance.

Costs and Scheduling

Repair and their design cost aircraft downtime and operating expenses. However, it is well established that it is cheaper to repair than replace, given appropriate facilities and adequate personnel skilled to do the repair.

Low Observables

A most important vehicle attribute for today's military is the reduced radar cross-section (RCS), which is a major contributor to its stealth characteristics. If the vehicle is a stealth type of design, then repairs must be designed to maintain the mould line and not have reflective corners. In vehicles that have low stealth characteristics like the F-16 and F-18 aircraft types, stealth repair design is not a concern.

REPAIR OF MATRIX CRACKS

Matrix cracks are resin fractures within a ply of the laminate. We recall that matrix cracks have little impact on the structural strength integrity but can cause local stiffness losses and thus instability problems under compressive loading. Hence, there are two types of repairs required for matrix cracks:

- If the matrix cracks are insignificant as a damage type on the structural integrity of the composite laminate and they are exposed to the surface, then only a filling/sealing type repair is warranted. This type of repair will ensure that moisture and other fluids are excluded from the damaged area.
- However, if damage analysis indicates that local structural instability is likely, then the damaged region is filled and sealed with a doubler patch to restore local stiffness (Figure 7.1). The restored stiffness should be equivalent to the undamaged parent laminate stiffness such that

$$E_{\text{restored}_1} = \frac{E_{\text{patch}_1} h}{t_{\text{repair}}} = \frac{E_{\text{patch}_1} t_{\text{patch}} + E_{\text{damage}_1} t_{\text{damage}}}{t_{\text{repair}}}$$

$$= \text{Effective Patch Stiffness} + \text{Effective Degraded Stiffness}$$

where
$\quad E_{\text{parent}_1}$ = parent laminate effective principle stiffness
$\quad\quad h$ = laminate thickness

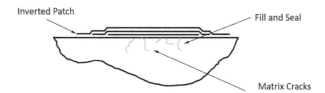

Inverted Patch   Fill and Seal

Matrix Cracks

FIGURE 7.1   Doubler cover patch over surface matrix cracks.

t_{repair} = thickness of repaired region
E_{patch_1} = patch stiffness
t_{patch} = patch thickness
E_{damage_1} = damaged region stiffness
t_{damage} = depth of damaged region

A simpler method of determining the load sharing ratio with a bonded doubler repair is illustrated in Table 7.2.

TABLE 7.2

Simplified Method for Analysis and Strip Analysis of Load-Sharing with Bonded Repairs

Simplified Method for Analysis	Strip Analysis

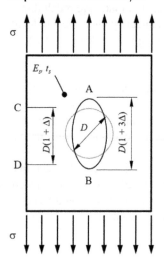

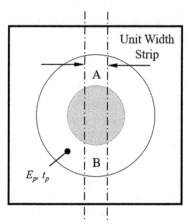

Stiffness from C to D	Stiffness of patch from A to B
$$k = \frac{P}{\Delta}$$	$$k = \frac{E_p t_p}{D}$$
$$k = \frac{E_s t_s}{D}$$	
Stiffness from A to B	Load is shared roughly in ratio of relative stiffnesses, so load passing through unit strip of bond shown is:
$$k = \frac{E_s t_s}{3D}$$	$$\frac{1}{1 + \left(\dfrac{E_s t_s}{3 E_p t_p} \right)} \quad \text{of total}$$
a. Behaviour of Unrepaired Skin with Hole	b. Load Sharing after Patch Bonded On

In both matrix crack repair scheme designs:

• The damage is not cut-out, because load carrying fibres are still in place.
• Prior to installing the repair scheme, the local damaged region is dried out.
• A low viscous filling/sealing resin is used to fill the surface cracks.

As good design practice, doubler patches are stacked in reverse order with the largest ply on the top (Figure 7.1). This practice provides both a final seal to the repaired area and reduces peel stresses due to more uniform load transferred through the repair patch.

REPAIR OF DELAMINATIONS

Delaminations are more of a concern to the structural stability of composite laminates rather than strength degradation; hence, only a fill/doubler patch repair is normally warranted. If the delamination effect on the structural integrity is within the damage tolerant design strain allowables, then no repair is required; however, if the delamination is exposed to a free edge, then filling with a low viscous resin and sealing is necessary to prevent environmental degradation. The structural repair of internal and edge delaminations are as follows:

INTERNAL DELAMINATIONS

Internal delaminations under compressive loading, that are close to the surface, and that the damage analysis shows that the sub-laminate is likely to buckle under design allowable strains require a doubler patch to increase the local stiffness of the sub-laminate (Figure 7.2).

Determination of the patch stiffness is based on the analysis above, but here, the stiffness of the sub-laminate and patch needs to be such that the critical buckling load is greater than the applied design allowable load. Or, more simply, ensure that the stiffness of the patch is equivalent to the undamaged laminate stiffness, such that

$$E_{\text{patch}} = \frac{E_{\text{parent}} t_{\text{parent}}}{t_{\text{repair}}}$$

FREE EDGE DELAMINATION

With an edge delamination, the first requirement of the repair is to seal the edge from further moisture absorption; again, a low viscous resin is used. Local stiffening of the edge is more difficult since the driving out-of-plane forces are still present. The most effective repair design is to simply reinforce the out-of-plane direction. Since the out-of-plane stresses are much lower than in-plane, a fastener or thin capping patch is all that is required (Figure 7.3).

If the delaminations are severe, then the damaged region will have to be removed, and the repair scheme will then be for that of a hole.

Prior to repair scheme installation, the damage region will require the moisture to be removed.

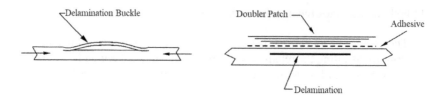

FIGURE 7.2   Doubler patch installation over delamination.

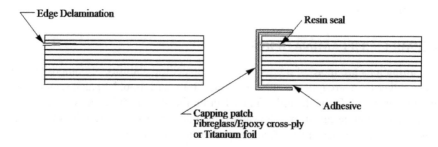

FIGURE 7.3   Out-of-plane reinforcing capping patch over edge delamination.

Current practice in the restoration of local stiffness in the presence of a delamination is to drill two holes, where one is used to inject a low viscous resin and the other hole acts as a vent (Figure 7.4). This repair method, although very simple, has several drawbacks.

These drawbacks are listed as:

- The internal delamination is now exposed to the atmosphere.
- Moisture can be more readily absorbed into the delamination region via the resin plug.
- Injection of the resin can induce swelling forces that could induce delamination crack growth.
- Finding the ply levels to drill to where the delaminations are can be very difficult.
- By drilling the hole, a stress concentration is induced into the laminate.
- Often, the resin only partially fills the delaminated area.

REPAIR OF DEBONDS

A debond in an adhesively bonded joint can be located at the free edge, close to the free edge or centrally located (Figures 7.5 through 7.7). Repair assessment will identify that the repair will be semi-structural or structural. A semi-structural repair will be a doubler type repair, and a structural repair will require removal of the debonded region and the splicing of the repair patch with a new adhesive bondline created.

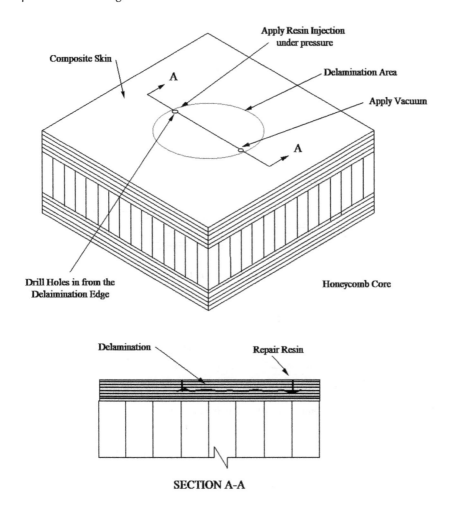

FIGURE 7.4   Ineffective repair of internal delaminations.

There are two regions of debonds which require specific repair design (semi-structural or structural). These two regions of interest are either:

• An internal debond
• An edge debond

INTERNAL DEBONDS

Under tensile loading, the internal debond must be large to warrant structural repair, whereas under compressive loading, the structural instability must be rectified by a bonded doubler patch. Rapid repairs of an internal debonded region involved fastener installation (Figure 7.8). For a large debond under tension, Figure 7.9 illustrates the

FIGURE 7.5   Edge debond in an adhesively bonded joint.

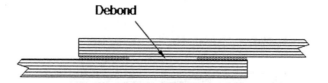

FIGURE 7.6   Internal edge debond in an adhesively bonded joint.

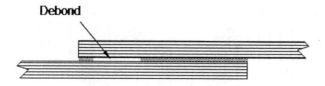

FIGURE 7.7   Central debond in an adhesively bonded joint.

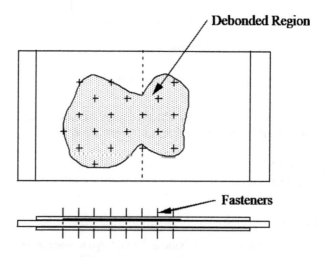

FIGURE 7.8   Rapid repair of debonded region.

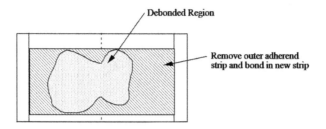

FIGURE 7.9   Large debond repair for tensile loading.

typical repair. As shown in Figure 7.9, the damaged region material is removed, and a flush patch is bonded. If the debonded region is not removed, a doubler patch over the damaged region is installed (Figure 7.10). The doubler patch is used to reduce the debonded outer doubler from buckling. To restore the effective stiffness of a debonded region under compressive loading, combined flexural stiffness of the outer adherend and doubler patch is calculated and compared against the compressive loads that caused buckling of the damaged region in the first place. The method of determining the thickness of the bonded doubler against compression buckling is the same as in delamination repair design.

FREE EDGE DEBONDS

The first requirement of a repair to an free-edge debond is to seal the damage from further environmental degradation. Prior to repairing a free-edge debond, the region must be dried out. Most repair approaches to the grind-out the debonded sub-laminate and repair with a scarf patch. Such a repair scheme damages the fibres that can reduce structural integrity more than what the structural integrity was with just the free-edge debond. An alternative repair scheme is to bond a doubler patch over the debonded regions and free-edge. This alternative repair scheme provide stiffness restoration without loss of fibre structural performance of the original debonded sub-laminate. Now, the main problem is to reinforce the doubler repair patch against induced peel stresses and subsequent debond growth under cyclic loading. To aid in sealing of the free-edge crack a low viscous resin is infused into the crack exposed free-edge. After resin sealing, an overlapping free-edge doubler is bonded in place (Figure 7.11).

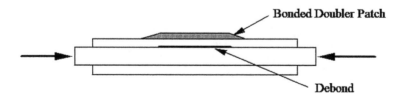

FIGURE 7.10   Bonded doubler patch repair for compressive loaded debonded region.

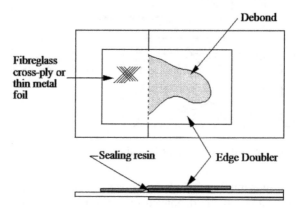

FIGURE 7.11   Edge doubler patch.

REPAIR OF HOLES AND FIBRE FRACTURE

LOW STRENGTH DEGRADING HOLES

Where the hole in the laminate presents a low strength degradation to the overall laminate structural integrity, the general repair is a plug/patch scheme, as shown in Figure 5.6. For a rapid-type repair, the damage is not necessarily removed, but in the majority of cases, the damage will be removed. The amount of damage to be removed should be minimal and of simple geometric design, i.e. a circle or dome-ended rectangle (Figures 7.12 and 7.13). The plug and patch should be of a lower modulus than the parent material so that the repair area does not attract load. With a low modulus plug, the stress concentration factor will reduce due to hole deformation constraints. The patch is mainly used as a sealing cover over the plugged hole.

MODERATE-STRENGTH DEGRADING HOLES

In the case where the damage analysis of a hole in a composite laminate indicates that there is moderate strength degradation, i.e. the current level of damage tolerance is significantly reduced, but catastrophic failure would only occur with severe

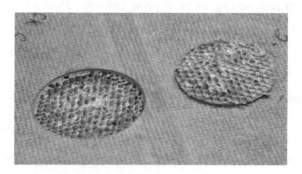

FIGURE 7.12   Damaged removed as a circle.

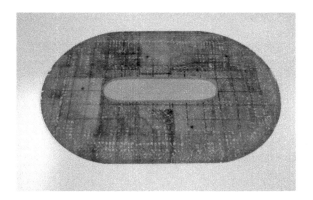

FIGURE 7.13 Damaged removed as a dome ended rectangle.

overload, then a low modulus plug and structural doubler patch are recommended (Figures 7.14 and 7.15). Again, the plug is of low modulus so that the load path is from the parent laminate into the doubler but not into the plug.

The design of the patch follows a simple method, which meets the following conditions:

- Half a double-lap joint is designed, acknowledging supports for bending resistance.
- A tapered patch is used to reduce peel, particularly when the thickness of the patch is greater than 1 mm (8 plies).
- The patch stiffness and thermal expansion coefficients are matched.
- The hole is not tapered.

Based on the idealised adhesive stress/strain curve (Figure 7.16), the load carrying capacity of the joint is defined as

$$P_{\text{adhesive\_shear}} = 2\sqrt{t_{\text{adhesive}}\tau_p\left(\frac{\gamma_e}{2}+\gamma_p\right)(Et)_{\text{patch}}}$$

where

t_{adhesive} = adhesive thickness (nominally 0.125–0.250 mm)
$(Et)_{\text{patch}}$ = effective stiffness of the patch or parent laminate

The allowable load per unit width of the patch is given by

$$P_{\text{all}} = Et\varepsilon_{\text{all}}$$

where

$$\varepsilon_{\text{all}} = 2{,}677\ \mu\text{strain}$$

$$= \varepsilon_{\text{limit}}/\text{Factor of Safety}$$

$$= \frac{4{,}000}{1.5}$$

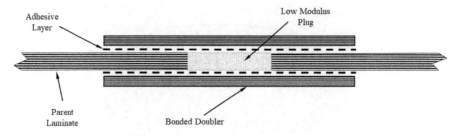

FIGURE 7.14   Thin laminate doubler repair patch.

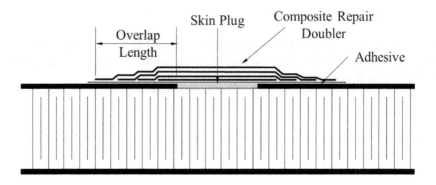

FIGURE 7.15   Structural doubler and plug repair in a sandwich structure.

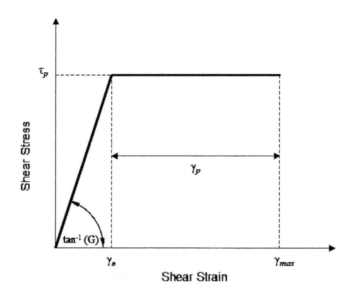

FIGURE 7.16   Idealised adhesive stress/strain curve.

The patch overlap length is

$$l_{\text{overlap}} = \left[\frac{P_{\text{all}}}{\tau_p} + \frac{2}{\lambda} \right] FS$$

where

$$\lambda = \sqrt{\frac{2G}{t_{\text{adhesive}} (Et)_{\text{patch}}}}$$

$$G = \frac{\tau_p}{\gamma_e}$$

FS = Factor of Safety on the adhesive bondline length

FULL STRUCTURAL REPAIRS OF HOLES

When the hole causes a significant reduction to the laminate strength, a fully structural restoring repair is required. The repair will either be a scarf (or stepped-lap) bonded patch for thinner structures, or for thicker sections, a bolted patch is recommended.

Thin Section Flush Repairs

A flush repair to a thin laminated section will either be a scarf joint or stepped-lap. Generally, the damage removed results in a sloping or scarfed hole. However, the repair ply configuration is actually a stepped arrangement (Figure 7.17). Using a simple scarf joint analysis approach, the final repair design will suit both bonded joint repair configurations. In the simple analysis, we try to maintain stiffness balance $(Et)_{\text{parent}} = (Et)_{\text{patch}}$, and matched thermal coefficient of expansion ($\alpha_{\text{parent}} = \alpha_{\text{patch}}$). Therefore, if the load is acting over a scarf angle of $\theta°$, the normal and shear stresses are (Figure 7.18 illustrates the geometry of the repair scheme)

$$\tau_{\text{adhesive}} = \frac{P\sin(2\theta)}{2t}$$

$$\sigma_{\text{adhesive}} = \frac{P\sin^2(\theta)}{t}$$

As θ gets smaller, the normal stress in the adhesive (σ) will approach zero. By letting the adhesive shear stress equal the adhesive plastic shear strength ($\tau = \tau_p$), the allowable adhesive load per unit width is given by

FIGURE 7.17   Scarf joint of original structure and stepped plies for the repair laminate.

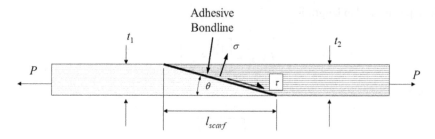

FIGURE 7.18   Scarf joint analysis geometry.

$$P = E\varepsilon_{all}t = \frac{2\tau_p t}{\sin(2\theta)}$$

Therefore, the scarf taper angle is

$$\theta = \frac{\tau_p}{E\varepsilon_{all}} \quad \text{in radians}$$

For the majority of structural adhesives typically used in the aerospace industry and the adherends that connect them, the scarf angle is limited for $\theta \leq 3°$. This gives a repair scheme length expression of (Figure 7.19)

$$l_{patch} = \frac{2t}{\tan\theta} + D_{hole}$$

where
$\quad\quad t$ = laminate thickness
$\quad\quad D_{hole}$ = the hole diameter

As a function of the laminate thickness for a hole size of 50 mm, Table 7.3 illustrates the increasing patch size if a scarf joint is used and clearly shows that this repair design scheme is limited to the repair of thinner sections.

Thick Section Bolted Repairs
First, we ask the question, "What defines a thick section laminate?" If a laminate is greater than 19–20 mm (3/4″) it is generally considered a laminate that is basically

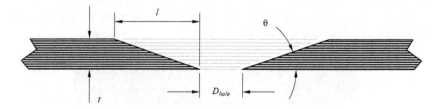

FIGURE 7.19   Scarf repair length geometry.

TABLE 7.3

A 3-Degree Scarf Patch Size-to-Laminate Thickness for a 50 mm Repair Hole

# of Plies	Laminate Thickness (mm)	Scarf Length (mm)	Total Patch Length (mm)	l_{patch}/D_{hole}
8	1.0	19.1	88.2	1.8
12	1.5	28.6	107.3	2.2
16	2.0	38.2	126.4	2.5
24	3.0	57.3	164.6	3.3
36	4.5	85.9	221.9	4.4
50	6.25	119.4	288.7	5.8

thick. Thick composite laminates are typically repaired with a bolted patch rather than a bonded scarf or doubler patch repair. In thick composite laminates a bonded repair patch can be geometrically inefficient (the bonded repair patch is overly large in length). Thus, a bolted repair is more practical and geometrically efficient, i.e. the overall size of the repair scheme is relatively smaller. The analysis of a bolted repair follows the methodology of a mechanically fastened joint using the allowable load per unit width of the parent laminate and repair patch is detailed in Chapter 6 of this book.

With reference to Figure 7.20, the following design points are recommended for a bolted repair design in composite laminates:

- Generally, use a low modulus plug that fills the hole of the damaged laminate. The low modulus plug restricts the load being transferred into the filled hole; a loaded filler hole has a greater strength reduction than an unloaded filled hole.
- Where a repair patch is relatively thick, taper the edges (Figure 7.20) or use a stepped-lap configuration (Figure 7.21). This approach will reduce the

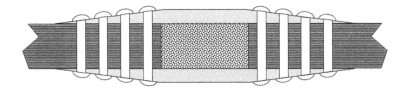

FIGURE 7.20   Typical cross-section of a tapered bolted repair patch on a composite laminate.

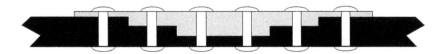

FIGURE 7.21   Bolted stepped-lap repair scheme on a composite laminate.

load transferred through the first row of fasteners. The first row of fasteners is usually the critically loaded row in mechanically fastened joints (see Chapter 6).

- Seal the repair patch to the parent laminate and install the fasteners with sealant in the wet condition, as this will aid in corrosion prevention.

SUMMARY OF REPAIR SCHEMES

The design of the composite repair scheme is based on many factors. These factors will influence the type of repair to be designed and installed. The engineering analysis of the damage stress state will identify repair types based on the level of structural integrity required to be restored. The basic repair schemes are either a cosmetic repair (non-structural), semi-structural repair (damage tolerance restoration), or a structural repair (restoring structural integrity). Within each of these repair schemes, the design approach will vary on whether an adhesively bonded repair or a bolted repair scheme is considered and if the structure is monolithic or a sandwich structure.

Throughout the repair scheme design and development, two important aspects are kept in mind. The first is to maintain a relative stiffness balance between the parent material and that of the repair patch, i.e. $E_{parent}t_{parent} = E_{repair}t_{repair}$. The second aspect is to maintain the thermal balance between the parent structure and the repair patch ($\alpha_{parent} = \alpha_{repair}$). These two aspects will assist in providing higher structural efficiency in the repair.

8 Repair Scheme Application

INTRODUCTION

Composite repair scheme application requires a number of fundamental steps. The first is to remove the already identified damage from the parent structure. A fundamental requirement when undertaking a heat-cured repair is the removal of entrapped water and absorbed moisture in the parent composite structure. With any repair process, there is a requirement to undertake some form of surface preparation of both the structure to be repaired and the repair scheme (patch). The repair patch is prepared to the specifications of the repair design. Once the repair scheme is fabricated, it is then installed onto the damaged removed area of the parent composite structure. Finally, the repair scheme is consolidated (cured) in place. These six steps are illustrated in Figure 8.1.

Post-repair inspection will follow the consolidation step. To assist in the repair application process, a set of instructions should always be written and reviewed prior to undertaking the repair scheme application.

DAMAGE REMOVAL

The removal of the damaged region is a compromised problem. The aim of the damage removal process is to remove only the damage materials but, at the same time, to create damaged removed shapes that are simple to repair and provide appropriate restoration of structural properties and performance. The latter requirement tends to lead to much more good material that is removed. However, for the sake of repair installation ease and improved structural performance of the repair scheme, the removed damage encompasses a significant proportion of sound material removed. Simple geometric damage removal patterns are best.

For plug and patch repair schemes, circular or dome-ended rectangle cutouts are suggested. Figures 8.2, 8.3 and 8.4 illustrate the best damage-removal patterns for plug/patch repair schemes.

The rounded-corner rectangle damage removed shows the least amount of good material removed in this particular case, but each damage removed pattern should be examined in each case. It is always good practice to keep the damaged removed geometry axes of symmetry aligned with the structural or orthotropic axes of the component. This practice will ensure that the repair patch plies are cut to the correct fibre pattern but remove the potential for fibre orientation errors.

The removal of damage for a scarf repair scheme in composite structure is based on the scarf angle required. The scarf angle is defined by joint design loads and/or

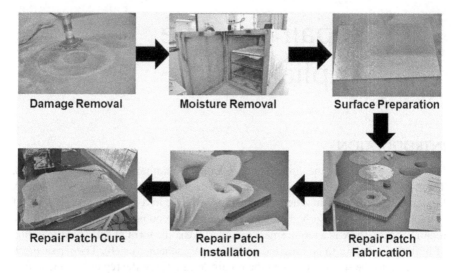

FIGURE 8.1 Composite repair scheme application steps.

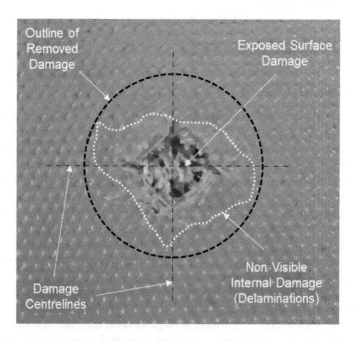

FIGURE 8.2 Circular damage removed pattern.

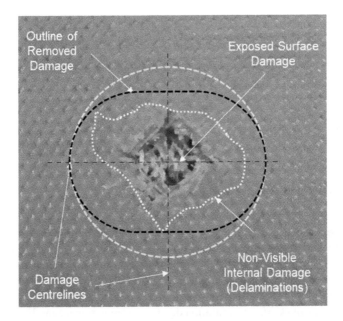

FIGURE 8.3 Domed end rectangle damage removed pattern.

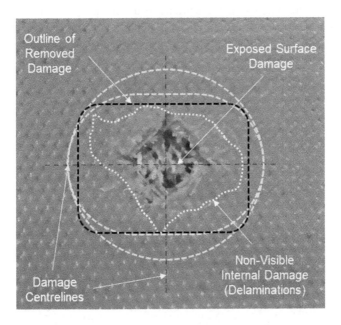

FIGURE 8.4 Rounded cornered rectangle damage removed pattern.

structural clearances. The typical scarf angle is between 1 and 3 degrees. For composite laminated structures, these angles equate to (Figure 8.5):

- For a 0.127 mm (0.005″) ply thickness of unidirectional lamina:
  - 6.35 mm (¼″) per ply … 1 degree or 1 in 50 slope
  - 3.175 mm (⅛th) inch per ply … 2.3 degrees or 1 in 25 slope
- For a 0.28 mm (0.011″) ply thickness of woven cloth lamina:
  - 6.35 mm (¼″) per ply … 2.5 degree or 1 in 23 slope
  - 3.175 mm (⅛th) inch per ply … 5 degrees or 1 in 11.5 slope

MOISTURE REMOVAL

The removed on entrapped and absorbed moisture in composite and sandwich structures is essential when high-temperature cured repairs are undertaken. The importance of this step in the repair process is when the cure of the adhesive is at a high temperature. Curing the adhesive and repairing patch resins at temperatures where either the parent laminate has adsorbed moisture and/or the core materials have either absorbed moisture or standing water can create expanded steam. There are well-known incidents of skin/core separation due to the expanding steam. Figures 8.6 and 8.7 illustrate the effects of skin/core debonding under heat-expanded moisture in both foam and honeycomb core, respectively.

The blown core outcome is the extreme result of moisture trapped in the core. However, a more common result of moisture in the core is bondline and repair patch porosity. Bondline porosity will always occur with moisture in the core, as shown in

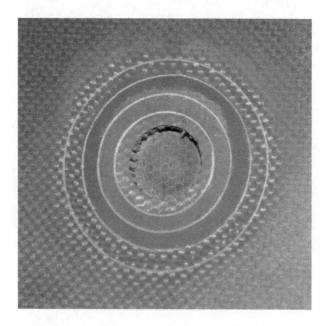

FIGURE 8.5   1/4 inch scarf angle.

FIGURE 8.6   Blown skin/core in Kevlar composite facing/foam core sandwich panel.

FIGURE 8.7   Blown skin/core in carbon composite facing/Nomex core sandwich panel.

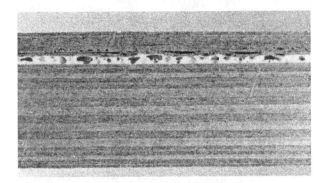

FIGURE 8.8   Bondline porosity due to moisture in the parent laminate.

Figure 8.8. Porosity in the repair patch will also occur if the composite repair patch is co-bonded in the curing process. Moisture degradation of the joint can be evaluated by either reducing the effective overlap length or reducing the effective adhesive properties. If the level of porosity is less than 25% by volume, a simple guide is to reduce the adhesive properties by the same percentage. Note that this reduction approach provides guidance rather than absolute structural integrity.

SURFACE PREPARATION

BONDING MECHANISM

The bonding mechanisms of interest to adhesively bonded repairs for composite materials are electrostatic, molecular, covalent and ionic bonding. Surface preparation of the region to be bonded is basically to increase the local surface energy and present a surface that will aid in good adhesive bonds that are durable. Figures 8.9 through 8.12 provide an illustration of the principle surface preparation outcomes.

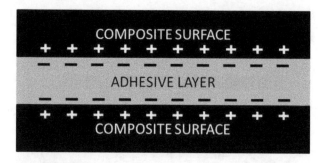

FIGURE 8.9 Electrostatic attraction (van der Waal forces).

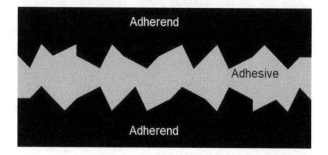

FIGURE 8.10 Mechanical locking through the surface roughness mechanism.

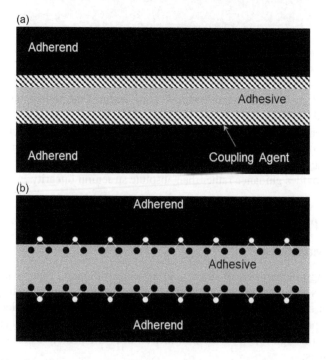

FIGURE 8.11 Covalent and ionic bonding mechanism.

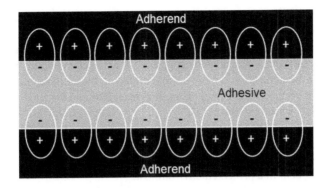

FIGURE 8.12   Molecular bonding mechanism.

Surface preparation for durability (long-life) of the adhesive bondline is slightly different for metals and composites. The following sections will discuss the surface preparation requirements for thermosetting resin composites, titanium and aluminium adherends.

BONDED JOINTS FOR COMPOSITES

Preparing the surface of a composite structure for adhesive bonding is relatively easy, but the principals of surface energy enhancement still apply. With most polymer resin-based composites, this is a form of degradation of the polymer surface from the environment. The breakdown of the polymer resin is a form of oxidation and leaves a soft, waxy residue. The surface preparation of a composite structure is based on the repair scheme considered as either a doubler repair or a scarf repair.

Composite Structure Surface Preparation for a Doubler Repair Scheme

The repair doubler is to be adhesively bonded to the existing surface of the composite structure. Typically, the damage region is removed to form a hole in the skin, and the doubler repair patch is bonded over the hole with an overlap length of 25 mm to 50 mm. The existing composite structure is to be cleaned and abraded for the durability of the adhesively bonded doubler repair patch. The recommended surface preparation of the original composite structure for a doubler repair scheme is as follows:

1. Ensure that hands are covered to protect the surface from natural oils, i.e. put gloves on.
2. Using a clean solvent and clean, lint-free cloth, wipe the Aluminium surface of interest in one direction and follow with a wipe of a dry, lint-free cloth in the same direction.
3. Repeat step (b) with a new and fresh, clean, lint-free cloth.
4. Abrade the surface with a Scotch-Brite pad (maroon colour preferred) with a distilled water solution. Abrade in a straight direction until the surface has a uniform pattern over the area to be bonded, plus 10 mm.

5. Remove the abrade residue with a clean, lint-free cloth with repeated one-direction wipes and replacing the cloth regularly.
6. Repeat step (d) with the abrade direction 90 degrees to that of step (d).
7. Remove the abrasion residue with a clean, lint-free cloth and distilled water with repeated one-direction wipes and replacing the cloth regularly.
8. Wipe the surface with a final clean, lint-free cloth and distilled water until no residue is seen on the cloth.
9. DO NOT touch the clean surface with anything except the adhesive to be used in the bonding process.
10. If adhesive curing is not immediate (within 30 minutes), place the composite component in a protected environment.
11. Undertake the adhesive bonding process within 8 hours.

Composite Structure Surface Preparation for a Scarf Repair Scheme

A scarf repair by definition will grind away the parent structure, leaving a bevelled edge from the removed damage to the top outer ply of the composite laminate. Since the parent material is machined (grinded), this leaves fresh and clean laminate that is ready to bond to. However, some minor surface cleaning is required to remove the grinding residue. The recommended surface preparation of the original composite structure for a scarf repair scheme is as follows:

1. Following the grinding operation, put gloves on.
2. Using a clean solvent and clean, lint-free cloth, wipe the composite surface of interest in one direction and follow with a wipe of a dry, lint-free cloth in the same direction.
3. Repeat step (b) with a new and fresh, clean, lint-free cloth and distilled water.
4. Dry the surface with clean compressed air.
5. DO NOT touch the clean surface with anything except the adhesive to be used in the bonding process.
6. If adhesive bonding is not immediate (within 30 minutes), place the composite component in a protected environment.
7. Undertake the adhesive bonding process within 8 hours.

BONDED JOINTS FOR METALS

The preparation of a metal surface for adhesive bonding is much more critical than that of a composite surface. The increase in criticality of surface preparation of a metal surface for adhesive bonding is due to the higher impact and rate of surface oxidisation. This process of surface oxidisation (known as hydration) initially produces a hard oxide layer (strong bond) with the exposed metal surface. Gradually, the oxidised surface becomes weaker and introduces a fracture path between the adhesive and the metal (an adhesion fracture). The basic requirement of the metal surface preparation process for metals is to clean, abrade, energise and enhance. The following sections provide the basis of the surface preparation for titanium patches and aluminium adherends.

Titanium Patch Surface Preparation for a Doubler Repair Scheme
A recommended procedure for the preparation of titanium patch materials in adhesive bonding is as follows:

1. Ensure that hands are covered to protect the surface from natural oils, i.e. put gloves on.
2. Using a clean solvent and clean, lint-free cloth, wipe the composite surface of interest in one direction and follow with a wipe of a dry, lint-free cloth in the same direction.
3. Repeat step (b) with a new and fresh, clean, lint-free cloth.
4. Abrade the surface with a Scotch-Brite pad (maroon colour preferred) with a distilled water solution. Abrade in a straight direction until the surface has a uniform pattern over the area to be bonded, plus 10 mm.
5. Remove the abrasion residue with a clean, lint-free cloth with repeated one-direction wipes and replacing the cloth regularly.
6. Repeat step (d) with the abrade direction 90 degrees to that of step (d).
7. Remove the abrasion residue with a clean, lint-free cloth and distilled water with repeated one-direction wipes, replacing the cloth regularly.
8. Wipe the surface with a final clean, lint-free cloth and distilled water until no residue is seen on the cloth.
9. Perform a surface chemical etch with hydrofluoric acid.
10. Apply a sol–gel surface adhesion promoter solution.
11. Apply a corrosion-inhibiting primer with a spray gun in a very thin, uniform layer.
12. DO NOT touch the clean surface with anything except the adhesive to be used in the bonding process.
13. Undertake the adhesive bonding process within 8 hours.

Surface preparation of titanium should be performed in a workshop and not in the field.

Aluminium Structure Surface Preparation for a Doubler Repair Scheme
A recommended procedure for the preparation of aluminium materials in adhesive bonding is as follows:

1. Ensure that hands are covered to protect the surface from natural oils, i.e. put gloves on.
2. Using a clean solvent and clean, lint-free cloth, wipe the composite surface of interest in one direction and follow with a wipe of a dry, lint-free cloth in the same direction.
3. Repeat step (b) with a new and fresh, clean, lint-free cloth.
4. Abrade the surface with a Scotch-Brite pad (maroon colour preferred) with a distilled water solution. Abrade in a straight direction until the surface has a uniform pattern over the area to be bonded, plus 10 mm.
5. Remove the abrasion residue with a clean, lint-free cloth with repeated one-direction wipes, replacing the cloth regularly.

6. Repeat step (d) with the abrade direction 90 degrees to that of step (d).
7. Remove the abrasion residue with a clean, lint-free cloth and distilled water with repeated one-direction wipes, replacing the cloth regularly.
8. Wipe the surface with a final clean, lint-free cloth and distilled water until no residue is seen on the cloth.
9. Highly recommended – Perform a light grit blast of the surface with an aluminium oxide non-recycled grit in a non-oil-based compressed air (nitrogen preferred) system.
10. Highly recommended – Perform a surface chemical etch with phosphoric acid.
11. Apply a sol–gel surface adhesion promoter solution.
12. Apply a corrosion-inhibiting primer with a spray gun in a very thin, uniform layer.
13. DO NOT touch the clean surface with anything except the adhesive to be used in the bonding process.
14. Undertake the adhesive bonding process within 8 hours.

Surface preparation of aluminium should be performed in a workshop and not in the field.

REPAIR SCHEME FABRICATION AND INSTALLATION

The fabrication of the repair patch and the installation of the repair scheme is not a trivial task. Several considerations must be understood and actioned, including the basic design configuration of the repair patch to be fabricated, the cure process based on the required repair scheme physical outcome and appearance, and cure monitoring requirements.

REPAIR SCHEME DESIGN

Further to the mechanical property requirements of the repair patch (discussed earlier in this book), several other aspects of the repair scheme must be considered and actioned (implemented or excluded). These other aspects cover the simplicity of the repair scheme, ease of installation, how best to achieve the repair scheme design requirements, honeycomb core ribbon alignment, the inclusion of the hole plug (for both monolithic and sandwich structure), the importance of repair pact alignment to the parent structural axis, and the repair patch ply stacking sequence.

Simplicity of the Repair Scheme

The simpler the repair scheme, the easier it is to fabricate and install. Simple repair scheme designs have a greater chance of being fabricated and installed and achieving the repair scheme design requirements. Simple repair schemes are quicker to fabricate and install and provide an opportunity to get the asset back into operational status.

Easier to Install. With clear and effective repair installation instruction, the composite technician can install the repair scheme relatively easily. An easily installed

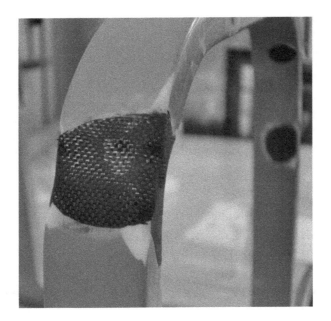

FIGURE 8.13 Repair installation made difficult due to location on the vehicle.

repair will also assist with the repair scheme meeting the repair design requirements. Note that not all repairs have gravity assisting. In a number of cases, the repair is a vertical or downward-facing surface (Figure 8.13).

Achieving Design Requirements

The simplicity and ease of installation of the repair scheme will provide a good chance for the repair scheme to meet the design requirements. Attention to detail during the repair scheme fabrication and installation will go a long way in producing a quality repair scheme.

Honeycomb Core Ribbon Alignment

In many small composite sandwich structures repairs, the positioning of core replacement plugs is not critical to the overall structural integrity of the component being repaired; however, the alignment of the core plug replacement for large repairs of sandwich structures is essential to the component structural integrity. So, the question is when honeycomb core ribbon direction is important. Since honeycomb core is an orthotropic material (there are directional property variations), then the best answer is to assume that the core alignment is always critical. Thus, best practice is to identify the parent structure ribbon direction and cut and install the honeycomb core replacement plug with the ribbon direction matching, as seen in Figure 8.14.

The Plug

The installation of a repair plug requires some important considerations. These considerations include how the repair plug will be held in place. Figure 8.14 shows how a

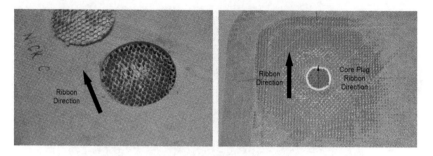

FIGURE 8.14   Honeycomb core repair plug alignment.

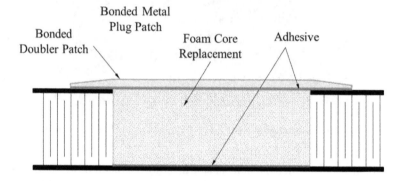

FIGURE 8.15   A plug-filled hole in a repair scheme.

foaming adhesive is used to attach the replacement core to the parent core. The bonding of the core to the parent inner skin will require an adhesive layer (Figure 8.15). Another very important requirement of the replacement plug is to consider if the parent loads are to be taken up by the plug. Typically, the plug does not require the axial load to be induced into it. To ensure this is the case, the plug stiffness must be at least one-tenth that of the parent structure:

$$\frac{\{Et\}_{\text{Parent}}}{\{Et\}_{\text{Plug}}} = \frac{\{Et\}_{\text{RepairPatch}}}{\{Et\}_{\text{Plug}}} > 10$$

Repair Patch Alignment

The alignment of the repair patch material axis is to be as close as possible to the structural axis of the parent structure (Figure 8.16). When using a quasi-isotropic lay-up, the alignment of the repair patch with the parent structure is not a major issue. As the parent structure increases in orthotropy, the repair patch alignment with that of the parent becomes increasingly critical. The degree of orthotropy can be estimated by

$$\text{DofO} = \frac{E_1}{\sqrt{E_2 G_{12}}}$$

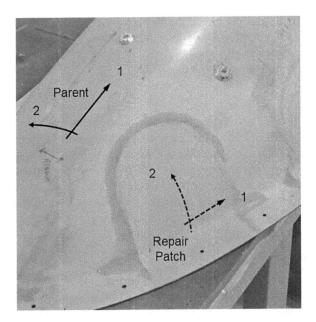

FIGURE 8.16   Repair patch alignment to parent structure.

Good engineering practice is to always have the repair patch align with the parent structure. This practice is achieved by using guidelines on both the repair patch and the parent structure to aid in alignment.

Repair Ply Stack on Parent

Basically, unless otherwise stated, the repair ply stacking will match the parent ply stacking. This action will maintain the stiffness balance between the repair scheme and parent structure ($E_r t_r = E_p t_p$). Note in Figure 8.17, for the doubler and scarf repair schemes, that the orientation and materials of the plies are equivalent, such that R1 = P1, R2 = P2, ... R8 = P8. In both repair schemes illustrated in Figure 8.17,

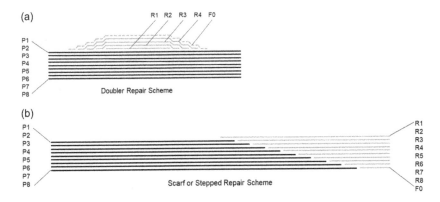

FIGURE 8.17   Repair stacking sequence recommendations.

ply F0 is used as a cover ply for the doubler repair scheme, and for the scarf repair scheme, F0 is a filler ply. The F0 cover and filler plies are best to be a low modulus configuration so as not to attract load.

The Cover Ply

The use of a cover ply is suggested by most OEMs, but why have a cover ply? A cover ply will provide an environmental seal at the repair patch edge and reduce repair patch edge interlaminar stresses. How should the cover ply be configured? If the cover ply is just an environmental sealing patch, then the patch design is of low modulus, so the cover ply does not attract load. If the cover ply is used in a modification of the repair patch modulus so as to be equivalent to the parent structure modulus, then the cover ply engineering properties will be important. Basically, the cover ply stiffness can be determined from:

ADHESIVE SELECTION

The selection of the adhesive was discussed earlier in this book. However, we are reminded of the importance of selecting the best or most appropriate adhesive for the repair. The following list states the things to consider when selecting an adhesive for the repair of composite structures:

- Compatibility with the parent and repair scheme resin system to ensure the best possible adhesion characteristics and durability
- Temperature of cure compatibility
- Environmental range (temperature, moisture, etc.) compatibility
- Application equipment and facility availability
- Curing parameters are acceptable
- Availability of the adhesive
- Cost of the adhesive and application equipment

REPAIR SCHEME CONSOLIDATION

CURE PROCESS

The cure process will include the following aspects: cure profile (temperature, pressure, vacuum with respect to time), the cure parameters (specifics of the cure profile), the bleed schedule (driving to the preferred fibre volume fraction and surface profile appearance), and other considerations. Each of the cure profile topics is explained in detail in the following paragraphs.

Cure Profile

The cure process for both a repair patch consolidation and adhesive bonding of a repair patch are based on the three parameters of temperature, vacuum and pressure against time. Figure 8.18 shows these three parameters as an ideal function of time. The actual values of the cure parameters for a particular resin system (composite and adhesive) are provided by the supplier of the system.

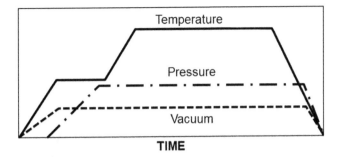

FIGURE 8.18 Cure parameters as a function of cure time.

Cure Parameters

The temperature parameter provides two functions for the repair cure. First, the resin system viscousity reduces as the temperature increases at the specified ramp-up rate (typically 1–4 degrees C per min). The low viscosity allows the resin to flow, thus providing higher uniformity of the resin/fibre distribution. The low viscosity flow also allows the entrapped air and volatiles to bleed off. The low viscosity state of the resin system provides the transportation qualities of the cure process. The second requirement of the temperature is to allow the resin system to solidify. The vacuum parameter in the cure process is to provide both lower void/porosity content in the repair patch and adhesive and provide thickness consolidation of the repair scheme. Whilst the vacuum parameter provides a level of pressure (at least one atmosphere), positive pressure can also be provided. For repair procedures, we find that non-autoclave cures are the norm. Non-autoclave pressures that are provided by mechanical force improve the thickness consistency of the repair scheme. Note that in Figure 8.18 that the external pressure is applied late in the cure profile. An example of a resin vendor cure profile is provided in Figure 8.19.

Bleed Schedule

The bleed schedule allows for repair scheme thickness consolidation, bleeding off entrapped air and volatiles, and achievement of the correct fibre volume ratio (repair patch mechanical properties). When repairing a composite structure, the structural properties have been finalised at the structure manufacturing stage. The laminated structure properties are typical at a high quality. The repair scheme properties will be finalised by the cure process used in repair. This might be different from the parent structure properties. The bleed schedule is used to produce a repair scheme with similar performance properties as the parent structure. The typical bleed schedule is illustrated in Figure 8.20.

Other Considerations

There are some other considerations that will aid in the success of consolidating the repair. These considerations include, but are not limited to, thermal couple placement, thermal profile mapping, vacuum bag leak check, pleating the vacuum bag, external thermal control, monitoring the cure with regular inspection, heater blanket

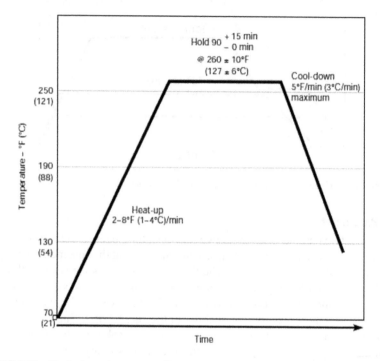

FIGURE 8.19   Typical vendor cure profile.

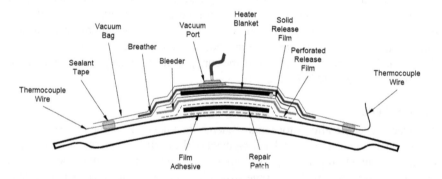

FIGURE 8.20   Typical (standard) bleed schedule for a composite repair patch.

vs. oven cure, using moulds, etc. Each of the above-mentioned cure consolidation considerations is now briefly described:

- **Thermal Couple Placement**. The parent structure might have areas that are thicker or thinner over the heated region. This could mean that there will be places on the repair that are heat sinks or higher temperature. Ideally, if there is a colder cure temperature on the repair, then the cure time should be increased. Typically, for every 5 degrees C below the optimal cure temperature, the cure time is increased by 20 minutes, but no more than 20 degrees

C maximum temperature drop. If there is a hot spot over the repair area, reduce the lower temperature limits on the heat controller and at cure time based on the comments earlier. An additional 30 minutes post-cure is also recommended in all cases.

- **Thermal Profile Mapping**. With complex internal structural arrangements of the parent structure being repaired, it is generally good practise to undertake a thermal survey to identify temperature range. Follow the suggestions for cure temperature variations and cure time as stated in the preceding sub-paragraph. If the temperature range is outside the range limits (i.e. +10 degrees C to −20 degrees C), then consider using two heater blankets and a dual-zone hot bonder controller.

- **Vacuum Bag Leak Check**. When a vacuum is applied to the repair scheme, a vacuum bag leak check is essential to ensure a high-quality repair (low levels of porosity in the repair patch and bondline). The vacuum bag leak check will show if there is or is not a vacuum bag leak. The vacuum bag leak check will not identify where the leak is, but typically, the leak will be along the vacuum sealing tape boundary, at the vacuum port if the structure being repaired is not a sealed surface (fastener holes or porosity), or simply, the vacuum bag integrity may have been compromised (there are pin holes in the vacuum bag). Figure 8.21 shows the use of a vacuum flow gauge that will identify vacuum bag leaks.

- **Pleating the Vacuum Bag**. If the composite repair is over a structure with moderate surface undulation or is a three-dimensional shape, then the vacuum bag must be pleated to ensure that there is no bridging of the vacuum bag. Vacuum bag bridging can cause vacuum bag rupture, poor repair patch consolidation and poor repair edge quality. Vacuum bag pleats with a repair are illustrated in Figure 8.22.

- **External Thermal Control**. The external environmental conditions (temperature and air flow) may create a surface-cooling condition of the repair scheme. Temperature control can be maintained with insulation blankets, as shown in Figure 8.23. If the insulation blankets do not achieve a satisfactory temperature control, then move the repair component to a better environment or cure the repair scheme in an oven, if practical.

- **Monitoring the Cure with Regular Inspection**. Never leave a composite repair cure process unattended for more than 15 minutes if out of audible range of any cure process alarm. Always set temperature and vacuum limits within reasonable ranges. There have been too many over-temperature repair processes that have caused more significant damage to the parent structure than was being originally repaired not to follow this advice.

- **Heater Blanket vs Oven Cure**. The correct and uniformity of the cure temperature is very important in the final outcome of a quality composite repair scheme. Heater blanket repair is common, particularly in structures in the field that are unable to removed. However, heater blankets do not give perfect thermal uniformity. Ovens provide a much better level of thermal uniformity, and if available, the use of an oven to cure a composite repair scheme is highly recommended.

- **Using Moulds**. Where surface profile and surface quality need to be maintained on the repair, a mould is required. A mould splash is taken of a pristine component surface, and when the repair is undertaken, the mould provides a flush and smooth repair patch outer surface. The outer surface mould line of the parent structure is thus maintained. Figure 8.24 illustrates a composite repair mould.

FIGURE 8.21   Vacuum bag leak detector – vacuum flow gauge.

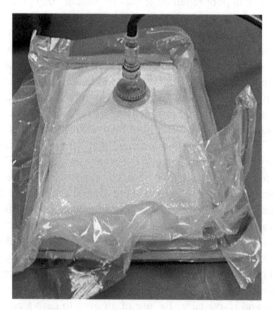

FIGURE 8.22   A pleated vacuum bag.

FIGURE 8.23   The use of an insulation blanket to better control the external environment.

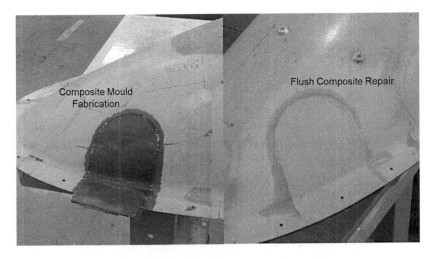

FIGURE 8.24   Composite repair mould fabrication and post-repair surface quality.

FASTENER CONSIDERATIONS

If the repair scheme requires bolts or rivets, several considerations in hole position, fastener spacing and edge distance, fastener pull-through limits, the laminate ply configuration implications on margins of safety, hole drilling and finishing require careful thought. Fasteners in composites have relatively low structural efficiencies and require more focus on the joint design and analysis.

POST-REPAIR INSPECTION

Following the removal of the curing medium (vacuum bagging and bleed schedule materials) and inspection of the repair scheme, it is essential to identify the quality and integrity of the repair scheme. Any of the typical NDI methods and techniques can be used in the post-repair inspection methods. Generally, we find that two approaches are most commonly used with two or three other more detailed methods used if necessary or for formal validation. The two most commonly used NDI methods are visual inspection followed by the simple tap test method. If the simple methods identify some concern in the repair scheme installation, then either or in combination ultrasonic inspection, x-ray and/or thermography NDI methods are used.

Visual Inspection

A visual observation of the repair scheme surface is used to identify an appropriate cure of the repair scheme, good surface appearance and proper resin flow. First, we inspect the quality and uniformity of the resin bleed and flow by observing the resin in the bleeder schedule (Figure 8.25). Then, closer inspection of the surface will reveal surface porosity (Figure 8.26).

Tap Testing

Lightly tapping the repair surface with an appropriate tool will allow identification of shallow sub-surface defects such as delaminations, debonds and large porosity or voids (Figure 8.27).

Ultrasonic Inspection

For more definitive examination of the internal integrity, ultrasonic inspection will aid in this identification. Figure 8.28 shows the application of an A-scan ultrasonic inspection process on a composite panel.

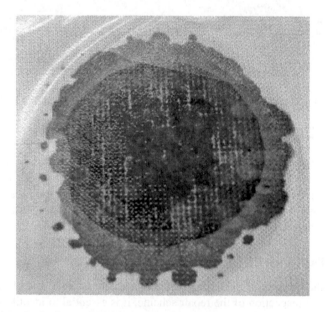

FIGURE 8.25   Resin flow pattern observed in the bleeder schedule.

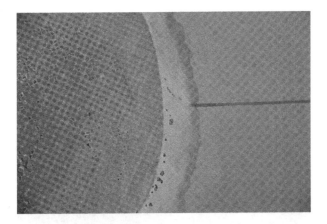

FIGURE 8.26   Repair scheme surface porosity.

FIGURE 8.27   Tap testing a panel.

X-Ray Inspection

For a repaired sandwich panel, the x-ray inspection provides useful information on the effect of internal resin content, for example the splicing of replacement core (Figure 8.29).

REPAIR INSTRUCTIONS

Introduction

The preparation and writing of composite repair instructions is essential in that it provides the composite repair technician with the details of how to physically do the repair. Without composite repair instructions, the best-designed repair schemes

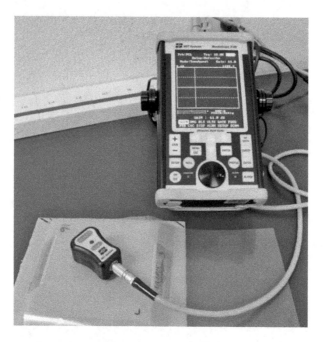

FIGURE 8.28   A-scanning of a composite sandwich panel.

FIGURE 8.29   X-ray of a sandwich panel repair core replacement.

have the potential to be implemented ineffectively or incorrectly. The careful drafting of the composite structures repair scheme should be done in collaboration with the composite repair technician. This collaboration will ensure that the composite repair technician understands what the composite repair design engineer is trying to achieve and that the composite repair engineer understands what the capabilities and foreseen difficulties explained by the composite repair technician are.

The composite repair preparation and installation instructions need to convey essential information to the composite repair technician apart from just the actual installation of the composite repair scheme and cure procedures. Safety and materials handling information, materials required and tools to be used in the repair, quality control and assurance measures should be explained, and post-repair documentation requirements must also be clearly specified. The incorporation of diagrams and illustrations into the composite repair instructions are essential in conveying the composite repair fabrication, preparation and installation needs and post-repair outcomes.

Generally, the composite repair instruction is written for the composite repair technician, who has had the appropriate training, knowledge and experience in composite repair, can understand. However, never assume that the composite repair technician is adequately trained or is expected to infer the intent of the composite repair instructions. Make sure that the composite repair instructions are complete and make reference to generic repair supporting instructions, such as surface preparation, so that there is no ambiguity in the requirements of the composite repair fabrication, preparation and installation.

CONTENT OF THE COMPOSITE REPAIR INSTRUCTION

Composite repair instruction must include the following topics to be a complete document:

- General introduction
- Safety and materials hazard information
- Reference documents
- Materials and consumables
- Tools and equipment
- Number of repair technicians required
- Time to complete the repair
- Repair documentation to be completed
- Quality control and quality assurance
- The repair instruction

General Introduction

The general introduction to a composite repair instruction simply outlines the aim of the composite repair scheme or what the repair instructions are going to specifically do and the basic content of the repair instruction. The length of the introduction should be no more than two paragraphs and be self-contained. For example:

This repair instruction will rectify typical damage to honeycomb core sandwich panel where the damage is a result of through penetration of both inner and outer skins and is away from free edges, core taper and internal structural members.

CONTENTS

Safety and Materials Hazard Information

The safety and materials hazard information provides a general interpretation of the words relating to *Notes*, *Cautions* and *Warning* banners, the hazards of the materials to be used and the appropriate safety equipment required. The actual *Notes*, *Cautions* and *Warnings* banner contents are always highlighted prior to the activity to which they are connected. Materials hazards are often referred to in the Safety Data Sheets (SDS), but Personal Protective Equipment (PPE) required should be listed.

Notes, *Cautions* and *Warnings* banners are identified by special fonts and/or borders to highlight them to the reader. For example:

NOTE

A Note describes or clarifies conditions or procedures.

CAUTIONS

A Caution describes conditions or procedures that could result in damage to or destruction of equipment if not followed correctly.

WARNING

A Warning describes conditions or procedures that could result in injury or death if not correctly followed.

Reference Documents

All documents that are quoted in the repair instructions are to be listed in this section of the composite repair instruction. The reference documents are usually other structural repair manuals (SRM) or general repair manuals, which are to be used in association with the composite repair instructions. Several of these references may also be generic company manuals and instructions.

Materials and Consumables

All the materials and consumables to be used in the repair are to be listed in this section. The listing of the materials and consumables will give the specification or part number and simple nomenclature that is used in the repair instruction. An estimate of the quantity to be used would also be most useful to the composite repair technician. If alternatives to materials and consumables are allowed or available, then they are listed here.

Specification or Part No.	Nomenclature	Quantity Required
MIL-S-83430	Sealing Compound	57 gms (2 oz)
FM 300 or FM 400	Film Adhesive	20 sq.cm (3 sq.in)

Tools and Equipment

All tools and equipment required to implement the composite repair must be listed. Alternatives may also be shown in the list.

Part/Tool Number	Nomenclature	Alternative
...XXX...	Tap Hammer	Coin
...YYY...	......	Not Applicable

Number of Repair Technicians Required

Depending on the complexity of the repair, more than one repair technician may be required to complete the composite repair. The minimum number of repair technicians should be clearly stated in this section. Also stated in this section is the level of skill required by the repair technician, particularly in safety critical structural composite repairs.

Time to Complete the Repair

To assist in the management of the composite repair facility and personnel requirements, an estimate of the time to complete the composite repair is recommended. This estimate has implications when a composite repair will take more than one shift to complete or carries over to the next day, particularly if certain steps in the repair must be completed in a set period of time.

Repair Documentation to be Completed

Documentation which is required to be written following the completion of the composite repair must be listed in this section. This documentation would include a composite repair check list and independent inspector's stamp column, shelf-lifed materials time-out chart, cure cycle chart, repair technician details, photographs of the repair process as required, repair installation temperature and humidity profiles, etc.

Quality Control and Quality Assurance

All the quality control requirements for the composite repair are to be specified in this section. A listing of the applicable quality assurance test methods to be employed should be given, including any special equipment and requirements. The actual details of each test are best discussed at the relevant repair step.

QUALITY CONTROL

Remove rolls of shelf-lifed materials from the freezer at least 12 hours prior to repair fabrication.

....

QUALITY ASSURANCE TESTS

Backing ply count during repair scheme fabrication.
<u>Coupon Tests</u>. A coupon of size 150 mm × 25 mm (6″ × 1″) using a $[0/\pm45/90]_s$ lay-up is to be manufactured concurrently with the repair.

....

THE REPAIR INSTRUCTIONS

The step-by-step repair instructions are described in detail. The detailed description should cover the following topics:

1. **Component Disassembly**. The disassembly of the composite component to be repaired will include a pre-inspection of the area to be repaired, safety of critical components and equipment, personnel safety and precautionary action required, control of removed components and equipment, and hazardous materials control measures to be in-place.
2. **Additional NDI**. Following disassembly of the composite component, additional NDI maybe called out when better access or a specific NDI process requires the component removal (i.e. radiography). This instruction section will detail the requirements of the additional NDI.
3. **Damage and Moisture Removal**. The removal of damage and moisture is explained in detail with appropriate control limits and cautions explained. Conditional limits of damage and moisture removal are also highlighted that may impact the engineering disposition and further assessment of the damaged area and repair approach.
4. **Repair Scheme Fabrication**. The details of the repair scheme fabrication, including materials required, size and shape of the repair scheme, configuration of the composite repair patch, debulking requirements and associated quality control requirements are explained. If precured patches are required, then the curing requirements are set-out in this section.
5. **Surface Preparation of Repair Scheme and Component**. Either a step-by-step process or reference to a specific support document for the preparation of the surfaces to be joined are listed in this section of the instruction. Both the component to be repaired and the repair scheme surface preparations are listed in this section.
6. **Repair Scheme Assembly and Application**. The assembly and application requirements and steps for the composite repair are detailed in this section. The positioning and limits of positional accuracy are explained

for the repair patch. Any challenging aspects to the repair application are explained, and advice for overcoming the challenges is provided.

7. **Curing or Fastening Requirements**. The curing requirements of a secondary bonded or co-bonded patch are detailed, including cure temperature steps, thermocouple positions, bleed schedule requirements and length of cure. Limitations for vacuum pressure allowable range and temperature variation allowable range are provided here. For a mechanically fastened pre-cured or metal repair patch, the fastener type, diameter and grip length are specified; drilling and reaming and the hole pattern are detailed.

8. **Post-Repair Clean-Up and NDI**. Clean-up requirements of the repair region, including surface conditioning (painting, etc.) are provided in this section. Any post-repair NDI action is also stated here.

9 Post-Repair Application Quality Assurance

INTRODUCTION

In any composite material and adhesively bonded structure repair workshop, one of the most important process requirements, and usually the one most often neglected, is **Quality Control and Quality Assurance**. In terms of a repair facility, quality control is related to the process that provides full acceptance by a customer of the integrity of the repaired product, the repair technology and the repair organisation. Without quality control requirements and quality assurance procedures in place, you (the repair organisation) risk the loss of your reputation and financial embarrassment due to poorly repaired products.

This chapter on quality control and quality assurance of repaired composite structures and components is aimed at providing the following information:

- That an understanding of composite and bonded structure repair workshop quality control is very important to the organisational success of the composite repair business
- The details of what material and process quality control requirements that are necessary in a repair workshop
- What are the typical quality assurance test methods to use that best suit your repair workshop?
- How do you make the best use of quality assurance test results?
- What are the guidelines for composite repair workshop practices?

This chapter provides the basic requirements of quality control and quality assurance for composite and bonded structure repair workshops. The implementation of these requirements will not be addressed in this chapter, as this topic should be addressed by company management. Implementation of quality control and its procedures are dependent on repair workshop fabrication requirements and repair levels, engineering and technician required skills, in-house control policy and procedures and testing equipment availability, end-customer needs and financial commitments of the company. Yes, quality control procedures and processes do not come cheaply – they do cost money, time and effort. These overheads must be absorbed into the repair service price. However, quality control and quality assurance costs pay dividends in bringing more customers and retaining existing customers. Quality control and quality assurance measures should be considered a revenue earner and not a debit in revenue. Much like advertising, quality control and quality assurance are money earners. Thus, the implementation of quality control policies and quality assurance measures will vary from one company to another. Excellent quality repair work does not come cheaply, but excellent quality repair work does not suffer from the cost of warranty re-work.

DEFINITION OF TERMS

The general term **quality control** can be separated into two distinct and independent components or actions. The first component of quality control is called **Material and Process Quality Control**, whilst the second component is in regard to **Product Quality Assurance Testing**. Each has an important place in total product quality outcome. The definitions of **Material and Process Quality Control** and **Product Quality Assurance Testing**, with respect to composite and bonded structure repair workshops, are detailed in the following paragraphs.

MATERIAL AND PROCESS QUALITY CONTROL

The **Material and Process Quality Control** aspects for composite and bonded structure repair workshops deals principally with the actual materials used in the fabrication of the component repair scheme, and the process by which the repair fabrication takes place. This particular component of quality control covers the entire product or service process. In addition, the level of skill in the workforce, both for technicians and engineers, is a fundamental part of **Material and Process Quality Control**.

PRODUCT QUALITY ASSURANCE TESTING

Product Quality Assurance Testing is the physical test of a product to show or demonstrate that it meets a set of known acceptance standards. This is to say, the product achieves the required level of mechanical strength, stiffness, durability properties, etc. for which it is designed. Quality assurance testing is the final proof of a successful quality control implementation.

THE ROLE AND IMPORTANCE OF QUALITY CONTROL

Generally speaking, the role of quality control in a composite and bonded structure repair workshop is to ensure that the product meets the stated requirements of the customer and/or certifying authority. The application of the role of quality control covers the entire production process, from material delivery through to end-product packaging and sale/commissioning. Additionally, today we also see quality control as an important requirement for the customer for the life of the product. Throughout the quality control process, whether it be component manufacture, major overhaul of the component or localised repair and maintenance, the procedures of quality control must be:

- Known by the engineering, technical and materials handling staff
- Stringently applied

Partial application of quality control procedures is as ineffective as doing nothing at all!

So, why is quality control so important? Again, the degree of importance placed on quality control is equated to the customer and/or regulatory authority requirements and the cost that both the producer and the customer are willing/required to pay. In the aerospace industry, both military and civilian, the importance of quality control is primarily derived from airworthiness standards and regulations.

With any repair action on composite and bonded structures, the materials, the processes and the end-product must strictly adhere to a defined specification conformance. Over the years, certain procedures for some component manufacture have relaxed, because they have exhibited consistent conformance to the specification. Only spot checks and product sampling occur once a product has shown consistent conformance to the specification. However, because of the relative infancy of composite and adhesively bonded structures industry, the product fabricator must demonstrate this conformance through rigidly following quality control procedures.

And finally, again, since the technology of repair of composite and adhesively bonded structures is relatively young, its application to critical structural components is new, and some past events have tarnished the acceptance of the repair of composite and adhesively bonded structures technology by a number of allied industries. Then, the importance of quality control in composite and adhesively bonded structure repair workshops should be utmost in the minds and attitudes of repair organisations.

MATERIAL AND PROCESS QUALITY CONTROL

To ensure that repair materials are of the highest quality and a repair station is set-up to maintain an appropriate working environment, certain quality control procedures or guidelines should be set in place. As a minimum, such guidelines would include the following:

- Capabilities to provide quality checks of supplied material
- Documented and controlled materials-handling procedures
- Provision of appropriate storage facilities for shelf-lifed materials
- Provision of an adequate clean room for repair fabrication and installation
- Provision of a separate damage removal area with suitable dust extraction equipment
- Installation of humidity and temperature controls in the clean room, including positive pressure for dust control
- Procedures for controlled access to the composite workshop
- Documentation of consumables life and storage, environmental controls and repair procedures followed
- Appropriately trained technical staff with set levels of skill requirements and on-going training programs

Each of these recommended practices is described in greater detail in the following sub-sections.

Supplied Materials Quality

A program of material quality checks through in-house or sub-contracted testing and evaluation is necessary when handling shelf-lifed high-strength composite and adhesive bonding materials. Such a program or policy would ensure quality materials are purchased and received. This requirement may not be necessary if the materials and equipment supplier has a proven and endorsed quality control program.

Materials Handling and Storage Facilities

All incoming materials need to be handled and stored in accordance with the manufacturer's materials data sheet and safety data sheet instructions. This is most necessary for hazardous and shelf-lifed materials, which are common in the composite and adhesive bonding industry. The cold storage of shelf-lifed materials, in particular the composite prepregs, film adhesives and one-part adhesives, is essential in order to maintain material property standards. The time in and out of cold storage must be well documented and the appropriate reduction in remaining life be made. An example of the calculation of out remaining useful life is given in the following example.

The thaw time of any materials held in cold storage is most important. When a material is taken from the freezer, it should not be removed from its protective, sealed wrapping until all of it has reached room temperature. If this procedure is not done, then water vapour will condense onto the surface and be absorbed into the resin materials. This moisture degradation will have the same effect on the mechanical properties of the cured material as using overaged materials, i.e. reduced mechanical properties. This thawing takes about 12 hours for large items, i.e. overnight. However, the degree to which moisture continues to condense on the sealing package can give an indication of the "ready to use" status of the material.

When the materials are returned into cold storage, ensure that the packaging is airtight, even to the point of removing the air by vacuum from the sealed package. This again will stop the moisture absorption potential whilst in cold storage. In general, storage temperatures −18°C (0°F), will provide the full storage life that is guaranteed by the materials supplier. Temperatures of −40°C (−40°F) will provide an indefinite life, but this is not guaranteed by the material suppliers. Finally, complete appropriate material storage documentation as required by company policy.

A roll of carbon/epoxy prepreg unidirectional tape is lifed at 12 months when stored at −18°C (0°F) and has 10 days usable life if stored at room temperature 25°C (75°F), as documented by the material vendor. If the roll is removed from the freezer, thawed for 12 hours and is used for a further 12 hours before being returned to the freezer, how much equivalent cold storage life has been used?

Let the maximum life be defined by T, then

The rate of degradation $(m) = \dfrac{1}{T}$

Hence, the equivalent cold storage time (t_{cs}) when the material is out the freezer for time t_o, is calculated from

$$t_{cs} = \left(\frac{m_o}{m_{cs}}\right)t_o, \quad \text{or}$$

$$t_{cs} = \left(\frac{T_{cs}}{T_o}\right)t_o$$

Therefore, for this example, $t_o = 1$ day (effective) 24 hours

Thus, $t_{cs} = \left(\dfrac{365}{10}\right)1 = 36.5$ days (say 37 days)

Thus, if the expiry date is, say, 15 January 2020, then the new expiry date should be 7 December 2019.

DAMAGE REMOVAL AND REPAIR FABRICATION ROOMS AND EQUIPMENT

There are two major processes undertaken in the repair of composite and bonded structures. These two processes require separation from a quality control point of view. These two major processes are:

- damage removal
- repair patch fabrication and application

Each of these processes must be done separately, in isolation of each other and with dedicated tooling for each process. The main reason for this separation is to stop the possible contamination of the fabricated repair scheme from contaminate-producing activities during repair. A certain amount of dust and grime is associated with damage removal, and if such contaminating produces are allowed to be introduced into the uncured prepregs or adhesives, then there is a likelihood of a reduction in the mechanical properties of the repair, which ultimately can degrade the repairs' ability to transfer load. Therefore, a two-room workshop is necessary in the overall repair of composite and adhesively bonded structures and components. A general composite repair workshop layout is shown in Figure 9.1.

The basic equipment requirements for both rooms are as follows:

- **Damage Removal Room**. The damage removal room (dirty room) major items of equipment are a high-efficiency dust-extracting system, with connections to both booth and portable vacuum hoses. The dust-extraction system is mainly for worker health and environmental safety but does remove the contaminant from the damaged component before the application of the repair scheme. All grinding, machining and cutting processes should be done in this room, as well as component tear-down and rebuild, where applicable. Also, the drying-out of damaged components is always necessary when undertaking heat-cured repair schemes, thus suitable equipment must be available. Any NDI process must be done in this room as coupling agents, etc. are known to contaminate prepregs, adhesives and bonded surfaces.

- **Clean Room**. In the clean room, or more specifically the fabrication room, the repair scheme is fabricated and applied to the cleaned damaged area. All the repair materials and consumables and cold storage equipment should be co-located in this room, with appropriately covered work benches and fume hoods for the preparation of resins and adhesives and curing equipment. Repair materials that do not require cold storage are stored on roll racks for easy access and in a non-contaminating manner (see Figure 9.2). The clean room must have some environmental control, as will be explained later, and this should also incorporate a level of positive pressure. This positive pressure will exclude dust from entering the clean room when a door is opened. There must be no machining or dust-producing activities carried out in the clean room.

Environmental Controls for Work Areas

In a composite and adhesive bonding workshop, the materials used are temperature- and moisture-sensitive. They are temperature-sensitive because the material fabrication process requires time and temperature to produce the finished product. A high working temperature can prematurely set resins and adhesives, thus shortening their pot life. Humidity control is important as polymeric materials absorb moisture

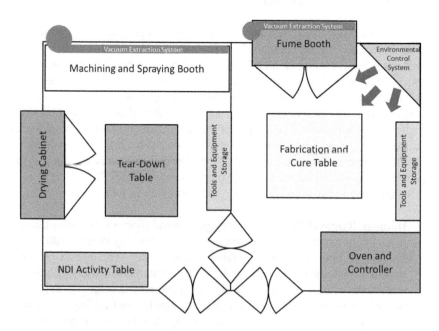

FIGURE 9.1 Typical workshop layout.

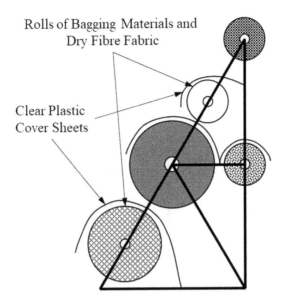

Rolls of Bagging Materials and
Dry Fibre Fabric

Clear Plastic
Cover Sheets

FIGURE 9.2   Consumable roll storage.

readily from the atmosphere. The desirable temperature and humidity conditions for composite work are shown in Figure 9.3, whilst environmental conditions for adhesive bonding are shown in Figure 9.4. In the repair of flight critical structures, environmental controls must not be waivered. If, however, field repairs are necessary, then the repair scheme is fabricated in a controlled environment (pre-cured patch), sealed and shipped to the field where it is attached as quickly as possible.

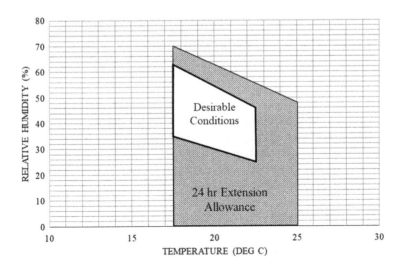

FIGURE 9.3   Composite workshop environmental requirements.

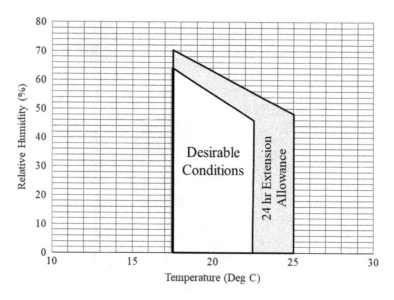

FIGURE 9.4   Adhesive bonding workshop recommended environmental conditions.

ACCESS CONTROL

The damage removal and clean rooms have materials that are hazardous to personnel and possess a contamination potential of components. Access to these rooms should be restricted to personnel trained to work therein. Indiscriminate access can cause hidden problems to personnel and repaired components. All personnel who enter must be appropriately clothed.

REPAIR DOCUMENTATION

In a quality control sense, there are four levels of repair documentation required to ensure that the repair will be of the highest standard. These repair documentation requirements are discussed in the following sub-paragraphs:

- **Equipment Operations**. The equipment used must be calibrated and continually checked. Specifically, fume booths and dust-extraction systems must be maintained to adequate standards for health reasons, the cold storage facility needs to maintain the required maximum temperature for full life of materials (i.e. temperature charts are necessary), and the clean room temperature and humidity levels must be monitored to ensure satisfactory levels.
- **Component Tracking**. The component to be repaired must carry with it the documentation showing the repair status and what materials and cure process have been and are to be used. The repair tracking documentation lists all of the vital information about the component, the repair scheme and sequence of repair action. Typically, the documentation should contain:

- Initial NDI survey
- Damage details
- Repair action and scheme
- Moisture removal requirements and method
- Damaged removal action
- Repair fabrication details, including materials used
- Repair application procedure
- Cure details
- Post-repair NDI action and results
- **Shelf-Lifed Materials Tracking**. Those materials that have a set shelf-life must have the appropriate documentation to show when they expire, including times out of storage and duration. Also, if the material has been re-lifed (requalified), then those details and the extended life must be shown.
- **Cure Documentation**. When the repair scheme is cured as a pre-cured patch or co-cured repair, the cure conditions should be documented. The details of this requirement should include the preferred temperature and pressure conditions and the achieved temperature, pressure and vacuum profiles for the entire cure process.

TRAINED PERSONNEL

Trained personnel in the composite and adhesive bonding workshop is of the utmost necessity. Particularly in the aviation industry cured composite and bonded metallic patches required high levels of engineering design development and technician hand skills. Like aircraft welding, once the repair has been completed any non-conformities in material and component properties are hidden until loaded. Furthermore, errors can be expensive to fix.

All personnel must be *qualified* in the preparation and handling of repair materials and repair methods. The inspectors must have the same qualifications as well as ample experience. The quality of the repair to maintain structural integrity throughout the airframe life is directly related to the design talents and workmanship employed.

The following concepts in the training and maintenance of learnt skills should be critically examined for each workshop environment:

- **Apprenticeships**. Since the technology is relatively new, and for most, the terminology not well understood, trained personnel, whether engineers or technicians, should service a qualifying period. This apprenticeship would refine the skills learnt and resolve any misinterpretations. Apprentices would be continually assessed during their qualifying period.
- **Work Experience**. The skills in the composite and bonding workshop are labour-intensive and thus require continual application to maintain those skills. Hand skills are quickly lost if not practised continually.
- **Training Updates and Extension**. Trained technicians and engineers must be given the opportunities to continually update and extend their learnt skills. This is particularly important in the composite and adhesive

bonding industry as new materials, processes and equipment become available. The composites and adhesives industry involves rapidly changing technology.

- **Continual Re-Assessment**. With the emergence of new and improved methods in technology and, indeed, the difficulties in identifying process and manufacturing errors, as well as the costly nature of mistakes, technicians should be continually re-assessed on their skills and competency. This re-assessment can be simply to routinely manufacture test coupons and test them to destruction.

QUALITY ASSURANCE

The implementation of quality assurance testing procedures provides physical evidence of the repair's integrity, whether they are direct tests or comparative coupon tests. Frequent quality assurance tests are, in particular, necessary during the developmental stages of a facility's growth. The frequency of tests can be reduced once a proven quality is reached and has been achieved consistently. The types of quality assurance tests required falls under three main areas, those of:

- Material quality assurance
- Process quality assurance
- Component quality assurance

MATERIAL QUALITY ASSURANCE

Material quality assurance is, in effect, pre-fabrication testing of constituent materials. Such testing is used to screen incoming materials to ensure that they conform to the manufacturer's documented standards and that re-life of time-expired materials can also be checked against that standard. There are two fundamental tests that can be used on prepreg composites and adhesives (or any resin material); they are detailed as follows:

- **Composite Prepreg Tack Test**. The prepreg tack test is a subjective test method in that the degree of tackiness of the resin in the prepreg is assessed. The assessment should be done with the prepreg at room temperature, within the desirable operating environmental conditions. If the prepreg displays the ability to stick, but be removed from another prepreg sheet, then the prepreg is said to have passed the test. However, if the prepreg has lost its tackiness, it may be in an overaged state. The degree of over-ageing must be assessed from a more quantitative test, such as that described in the adhesive flow test.
- **Adhesive Flow Test**. To qualify adhesives or resins are still in life, a very simple and inexpensive test is to determine the amount of flow. A small circular disk of film adhesive or measured amount of resin is placed in a known position between two layers of Mylar type film. This assembly is heated, under vacuum, to a temperature for the full cure cycle. The increased diameter of

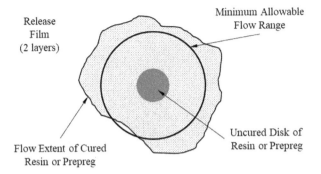

FIGURE 9.5   Typical flow pattern of a good resin/adhesive.

the flow is measured and compared with a known standard (Figure 9.5). If the flow is less than the standard, then it is reasonable to say that the material has overaged and should not be used in critical structural areas.

PROCESS QUALITY ASSURANCE

Co-fabrication quality assurance testing is that testing which takes place during the actual fabrication of the repair scheme and its placement on the damaged area. The testing, in effect, is simple in that little expense and time is required, but such tests have a large impact on the structural integrity of the final repair. The testing procedure also ensures that all repair steps are completed and that critical steps are independently inspected prior to continuing with the repair. The testing process is both a physical test and procedural checks. They are described as follows:

- **Backing Ply Count**. Each prepreg sheet comes with a backing ply. This backing ply remains with the prepreg through cutting and shaping and is removed prior to placement on the laminate stack. At the end of the lay-up procedure, the removed backing plies are counted and checked against the number of plies in the laminate. The counts should be the same, otherwise a backing ply is likely to be present in the laminate. This error is a typical location for delamination initiation.
- **Water Break Test**. The water break test is primarily for surface preparation checks in the bonding processes. The test can be used for both composite and metallic adherends. The principle of the test is that de-ionised water is sprayed onto the cleaned surface which is to be bonded. If the surface is completely wetted by the water, then it is ready for bonding. However, if the water congregates in puddles, then the surface is contaminated and requires further preparation. Figure 9.6 illustrates this principle well.
- **Repair Procedure Checklist**. Each and every repair instruction requires a checklist to be completed for each step. This will provide a degree of assurance against missed steps and assist the technician in accounting for the work completed. This can be very important when the repair continues

FIGURE 9.6   Good and poor water break test results. (Courtesy of John Hart-Smith.)

over several shifts and several technicians are involved. A simple example
of a repair procedure checklist is shown in Figure 9.7.

- **Independent Inspectors Stamp**. At the end of a critical set of steps, or
  a single major step, an independent inspector should validate the preceding steps with a stamp before the repair process can continue. This action
  would provide further assurance of repair integrity being of the highest
  standard. Figure 9.7 illustrates the independent inspector's stamp at the end
  of a step of critical repair process steps.

COMPONENT QUALITY ASSURANCE

Finished component quality assurance testing is a post-fabrication test procedure
which is comprised of either physical tests on the completed component for adequate

Step No.	Repair Procedure	Technician Check	Independent Inspectors Stamp
25	Abrade the surface with scoured soaked in solvent. Rub in one direction until a uniform surface appears, then rub perpendicular to that.	√	
26	Wipe off debris with a soft, lint-free cloth soaked in solvent. Change cloth regularly until the cloth shows no residue of contamination.	√	
27	Grit blast the surface as directed.	√	
28	Remove dust from the surface using a nitrogen, oil free air supply.	√	
29	Apply surface pre-conditioning treatment, as required.	√	
30	Conduct a water break test as directed.	√	X
31	Evaporate the water using a hot air drier	√	X

FIGURE 9.7   Repair procedure checklist.

material properties, i.e. a proof-load test or comparative coupon test, and/or a non-destructive test.

- **Physical Property Testing**. In physical testing, the proof-load test is very expensive, so the coupon comparative test is usually done. During the fabrication process, a small coupon of identical material is also fabricated. The fabrication process is also identical for the coupon. Following cure, the coupon is subjected to a range of physical tests to establish the material properties and compare them to manufacturer or analytically derived properties. The physical tests can include the following:
  - **Composites**. For composite coupons, four tests can be useful. They are as follows:
    - **Short Beam Shear Test**. The short beam shear test (Figure 9.8) is a good measure of the interlaminar shear strength, which is a matrix dominated property.
    - **Iosipescu Shear Test**. The Iosipescu shear test (Figure 9.9) measures shear strength and stiffness in a simple specimen. The Iosipescu shear test also measures the matrix materials.
    - **Tension and Compression Tests**. Simple tension and/or compression tests for general strength and stiffness properties. However, these tests tend to be fibre dominated and may not show the influence of degraded matrix material. A four-point beam test would probably be better.
  - **Bonded Joints**. There are two principal tests which can test the adhesive bond strength very simply. They are as follows:
    - **Lap-Shear Test, ASTM D 3165**. The lap-shear strength test (Figure 9.10) is for the determination of the static strength of a joint. Although the single lap joint has been designed to minimise peel stress, the failure in peel is still likely. However, if the test is to determine the strength of the adhesive, in that it has cured correctly, then the test, as a comparative method, is quite adequate.
    - **Boeing Wedge Test, ASTM D 3762**. The Boeing wedge test (Figure 9.11) is used to determine joint durability in a hostile environment. The test simply measures crack growth rate in the presence of a hot/wet environment and compares the rate of growth against a standard for that adhesive. This test will provide a degree of assurance of joint longevity.

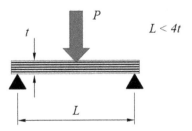

FIGURE 9.8 Short beam shear test.

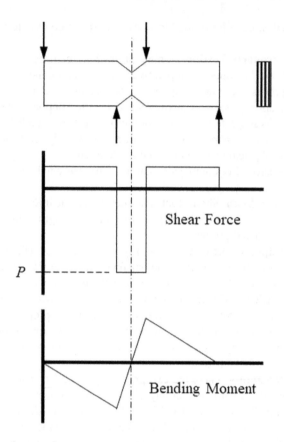

FIGURE 9.9 Iosipescu shear test specimen and loading.

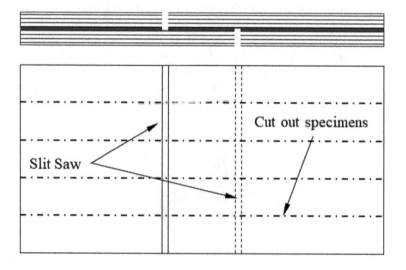

FIGURE 9.10 Lap-shear strength specimen – ASTM D3165.

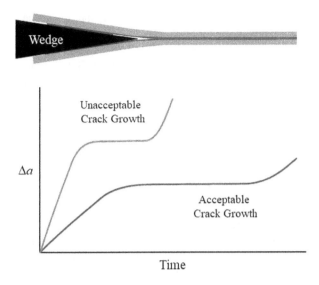

FIGURE 9.11   Boeing wedge specimen and test outcomes.

POST-REPAIR NDI

The final quality assurance test for composite and bonded structures, the final assurance of repair integrity, is to conduct a post-repair non-destructive test. Such tests range from very simple visual tests to the more exotic ultrasonic scanning. The requirement of NDI is to determine the presence of any non-conformity, in particular debonds and delaminations, porosity voids and foreign objects. A summary of post-repair NDI methods is shown in Table 9.1; however, the current practical NDI test methods are described from the simplest to the most difficult and expensive, but other accurate NDI methods are discussed in detail elsewhere.

- **Bondline Visual Inspection**. A bondline visual inspection will provide some assessment of the resin flow. This is more associated with bonded joints, but resin flow from cured laminates is also useful in identifying improper cure cycles or out-of-life resins. The typical resin flows from a bonded joint edge are shown in Figure 9.12.
- **Visual Inspection**. A pure visual inspection with the unaided eye can identify gross non-conformities such as bulges, warping and surface ply direction. The use of optical enhancing methods, i.e. a magnifying glass, can provide additional information.
- **Acoustic Resonance**. The simplest acoustic resonance method of NDI is the 'coin tap test', often referred to as the acoustic hammer test. This method has been used for many years and provides qualitative information of near sub-surface debonds or delaminations.
- **Ultrasonic Inspection**. The implementation of ultrasonic inspection is probably the most expensive of the methods currently available. The

methods can simply provide details of the depth and size of the non-conformity or full details of the topography of sub-surface defects.

- **New Methods**. Two new methods of full-field NDI that are rapidly showing great promise are shearography and holography. Both methods provide a visual, quantitative measure of sub-surface defect effects on the component under load. Whereas most other NDI methods indicate the size and shape of the non-conformity, these interferometric methods also show the component load effects; therefore, the criticality of the defect can be more easily assessed.

WORKSHOP PRACTICES

Considering the information provided on quality control and quality assurance, the development of specific required workshop practices in a composite and bonded structure repair facility will depend on the level and degree of repair, as well as the airworthiness authority's repair standards. Obviously, repair facilities that are required to perform structural repairs to highly loaded primary aircraft structures must follow all of the above, if not more, whereas repairs to lightly loaded structures, i.e. panels and secondary structures, are more relaxed in the repair process.

However, the aims and requirements of the repair must meet acceptable levels of structural integrity, and any repair, whether primary or cosmetic, must stay attached to the airframe, last for many years (durability) and cause no long-term effects to the remaining structure. Therefore, managers of repair facilities should carefully develop and implement appropriate quality control measures.

CONCLUSION

The process of quality control, which includes the quality of materials both incoming through to fabrication, the fabrication process and quality assurance testing, is an integral part of the entire component manufacture or repair process. Without quality control, the integrity of the finished product is virtually unknown, and in the aerospace industry, this is not only dangerous, but it is also totally intolerable.

Although the exact structural conformance of a component cannot be assessed without it being subjected to service conditions, the implementation and application of thorough quality control procedures will provide a good measure of the component's integrity. Furthermore, it provides both the supplier and customer a record of the component's integrity and highlights any trends that could lead to a serious loss of structural performance.

However, quality control does cost money. The implementation of quality control procedures must be at a level that either, or both, the supplier and/or customer can afford or, more importantly in the aerospace industry, what the regulatory authorities specify. But, without some quality control procedures in force, then, in the long term, both the product and the producer will suffer.

TABLE 9.1
Post-Repair NDI Methods

	NDI Method / Defect	Visual	Penetrant	Tap Testing	Bondtester	Pulse-Echo Ultrasonics	Through-Transmission Ultrasonics	X-ray	Dielectric	Thermography	Optic Interferometry	Microwave Absorption	Neutron Radiography	Mechanical Impedance
Laminate	Delaminations	1,2	1	√	√	√	√	3		√	√			√
	Macrocracks	1,2	2	√	√			3		√	√			
	Fibre Fracture							√		2,3	2,3			
	Interfacial Cracks									2,3	√			
	Microcracks		1	2	2					√	√			
	Porosity	1		2	2	√	√	√		2	√			
	Inclusions	1			2	2	2	√		√	√			√
	Heat Damage	1		2	2				2	2				
	Moisture							2	√	2		√	√	
	Voids				2	√	√	√		√	√			
	Surface Protrusions	√								√	√			
	Wrinkles	√								√	√			
	Improper Cure								√	2	2		√	
Bondline	Debonds	1,2	1	√	√	√	√	√		√	√			√
	Weak Bonds									2	√			
	Cracks	1,2	1	2	2	2	2	3		√	√			
	Voids			√	√	√	√	√		√	√			√
	Moisture							√	√		2	√	√	
	Inclusions			2	2	2	2	√		√	√			√
	Porosity				2	√	√	√		√	√			
	Lack of Adhesive			√	√	√	√	√		√	√			
Sandwich Panels	Blown Core			√	√	√	√	√					√	√
	Condensed Core			2	2		2	√					√	
	Crushed Cure			2	2		2	√					√	
	Distorted Core							√					√	
	Cut Core			√	√		√	√					√	
	Missing Core			2	2	2	2	√					√	√
	Node Debond							√		2	√			
	Water in Core			2	2		2	√				√	√	
	Debonds			√	√	√	√	√		√	√			√
	Voids			2	2	√	√	√		√	√			
	Core Filler Cracks			2		2	√	3		2	2			
	Lack of Filler			2	2	2	√	√		2	2		√	

Notes: 1. Open to surface.
2. Unreliable detection.
3. Orientation dependent.

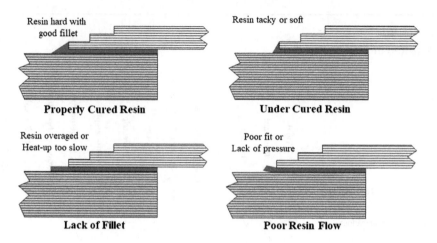

FIGURE 9.12   Bonded joint edge resin flow characteristics.

Bibliography

Adkins, D. W. and Pipes, R. B., Planar Scarf Joints in Composite Repair. ICCM/2, Proceedings of the Second International Conference on Composite Materials, Toronto, Canada, pp. 16–20 April 1978.

AFNOR, Non-Destructive Testing Standards Related to Composite Materials, Aero Datings, October 2013.

Agarwal, B. D. and Broutman, L. J., *Analysis and Performance of Fibre Composites*, 2nd Edition, John Wiley and Sons, New York, NY, 1990.

Ahmad, A. and Bond, L. J. (Eds.), *ASM Handbook, Volume 17: Nondestructive Evaluation and Quality Control*, ASM International, Russell, OH, 1989.

Ahn, S.-H. and Springer, G. S., Repair of Composite Laminates-I: Test Results, *Journal of Composite Materials*, 32, p. 1036, 1998.

Ahn, S.-H. and Springer, G. S., Repair of Composite Laminates-II: Models, *Journal of Composite Materials*, 32, p. 1076, 1998.

Ambartsumyan, S. A., *Theory of Anisotropic Plates-Strength, Stability, Vibration*, Volume 11 of the Progress in Materials Science Series, Ashton, J. E. (Ed.), Technomic Publishing Co., Inc., Lancaster PA, 1970.

Angell, G. and Markham, M. F., Characterization of Defects in Composites and Their Effect on Mechanical Strength, NPL Report, A(A)20, December 1980.

Antonuccia, V., Ricciardia, M. R., Caputob, F., Langellac, A., Loprestoc, V., Pagliarulod, V., Roccod, A. et al., Non-Destructive Techniques for the Impact Damage Investigation on Carbon Fibre Laminates, *Procedia Engineering*, 88, pp. 194–199, 2014.

Armstrong, K. B. and Barrett, R. T., *Care and Repair of Advanced Composites*, SAE, Warrendale, PA, 1998.

Armstrong, K. B., Bevan, G. and Cole, W., *Care and Repair of Advanced Composites*, 2nd Edition, SAE International, Warrendale, PA, 2005.

Ashton J.E. and Whitney J.M., *Theory of Laminated Plates*, Technomic, Lancaster, PA, 1970.

ASTM D3165-73, Standard Test Method for Strength Properties of Laminated Assemblies Adhesives in Shear by Tension Loading, 1979.

ASTM D3762-79, Standard Test Method for Adhesive – Bonded Surface Durability of Aluminium (Wedge Test), 1988.

ASM International, *Adhesives and Sealants, Engineered Materials Handbook*, Volume 3, ASM International, Russell, OH, 1990.

ASM International, *Composites, Engineered Materials Handbook*, Volume 1, ASM International, Russell, OH, May 1988.

Baker, A. A. and Jones, R. (Eds.), *Bonded Repair of Aircraft Structures*, Martinus Nijhoff Publ., Dordrecht, the Netherlands, 1988.

Baker, A. A., Rose, F. and Jones, R., *Advances in the Bonded Composite Repair of Metallic Aircraft Structure*, Elsevier Science, Oxford, UK, 2003.

Bar-Cohen, Y., Composite Materials – A Review, *Materials Evaluation*, 44, 4, pp. 446–454, 1986.

Bar-Cohen, Y. and Mal, A. K., Nondestructive Inspection and Quality Control – End-Product Nondestructive Evaluation of Adhesive-Bonded Composite Joints, Section 9, in *Adhesives and Sealants*, Volume 3, *Engineered Materials Handbook*, pp. 777–784, ASM International, Warrendale, PA, 1990.

Blitzer, T., *Honeycomb Technology*, Chapman and Hall, Melbourne, 1997.

Bruhn, E. F., *Analysis and Design of Flight Vehicle Structures*, Jacobs and Associates, Indianapolis, IN, 1973.

Chicken, S. H., Welburn, S. and Reed, S., Repairing Composite Aircraft Structures – RAF Experience of Peacetime and Battle Damage Techniques, *The Aeronautical Journal*, 277, June/July 1997.

Christensen, R. M., *Mechanics of Composite Materials*, J. Wiley and Son, New York, NY, 1979.

Chu, T. C., Ranson, W. F., Sutton, M. A. and Peters, W. H., Applications of Digital-Image-Correlation Techniques to Experimental Mechanics, *Experimental Mechanics*, 25, 3, pp. 232–244, 1985.

Chu, T., Mahajan, A. and Liu, C. T., An Economical Vision-Based Method to Obtain Whole-Field Deformation Profiles, *Experimental Techniques*, 26, 6, pp. 25–28, 2002.

Cloud, G. L., *Optical Methods of Engineering Analysis*, Cambridge University Press, New York, 1995.

Collings, T. A. and Beauchamp, M. J., Bearing Deflection Behaviour of a Loaded Hole in CFRP, *Composites*, 15, 1, p. 33, 1984.

Cook, B., Modeling and Experimental Measurement of Tensile Loaded Straight Scarf Joint, Master's Thesis, School of Engineering and Management, Air Force Institute of Technology, 2005.

Cytec Engineered Materials Inc., Technical Datasheet-FM 300 High-Shear Strength Modified Epoxy Film Adhesive, 2005.

Dally, J. W. and Riley, W. F., *Experimental Stress Analysis*, 3rd Edition, McGraw-Hill, Inc., New York, 1991.

Daniel, I. M. and Ishai, O., *Engineering Mechanics of Composite Materials*, Oxford University Press, New York, 1994.

Dorworth, L. C., Gardiner, G. L. and Mellema, G. M., *Essentials of Advanced Composite Fabrication and Repair*, Aviation Supplies and Academic, Newcastle, WA, 2010.

Dostal, C. A. (Ed.), *Adhesives and Sealants, Engineered Materials Handbook*, Volume 3, ASM International, Russell, OH, 1990.

Duong, C. N. and Wang, C. H, *Composite Repair: Theory and Design*, Elsevier, Amsterdam, The Netherlands, 2007.

Dutton, S., Kelly, D. and Baker, A., *Composite Materials for Aircraft Structures*, 2nd Edition, AIAA Education Series, Reston, VA, 2004.

ESDU, ESDU 82025 Amendment A, Failure Modes of Fibre Reinforced Laminates, June 1986.

Flabel, J.-C., *Practical Stress Analysis for Design Engineers*, Lake City Publ., Hayden Lake ID, 1997.

Frame, C. S., Composite Repair of Composite Structures, Paper 17, Proceedings of the 79th Meeting of the AGARD Structures and Materials Panel on Composite Repair of Military Aircraft Structures, Seville, Spain, 3–5 October 1994.

Frame, C. S. and Jackson, G., Defect Occurrences in the Manufacture of Large CFC Structures and Work Associated with Defects, Damage and Repair of CFC Components, in Chapter 21, *Characterization, Analysis and Significance of Defects in Composite Materials*, AGARD-CP-335, 1983.

Garrett, R. A., Effect of Defects on Aircraft Composite Structures, in Chapter 19, *Characterization, Analysis and Significance of Defects in Composite Materials*, AGARD-CP-335, 1983.

Garrett, R. A., Effect of Manufacturing Defects and Service-Induced Damage on the Strength of Aircraft Composite Structures, *Composite Materials – Testing and Design (7th Conference)*, ASTM STP 893, 5–33, 1986.

Good, G. W. and Nakagawara, V. B., Vision Standards and Testing Requirements for Nondestructive Inspection (NDI) and Testing (NDT) Personnel and Visual Inspectors, Aviation Maintenance Human Factors Program Manager, AAR-100, 2005.

Greszczuk, L. B., Consideration of Failure Modes in the Design of Composite Structures, in Failure Modes of Composite Materials with Organic Matrices and Their Consequences on Design, AGARD-CP-163, Paper No 12, October 1974.

Greszczuk, L. B., On Failure Modes of Unidirectional Composites Under Compressive Loading, Fracture of Composite Materials, Proceedings of the 2nd USA-USSR Symposium, Bethlehem, PA, pp. 231–244, March 1981.

Greszczuk, L. B., Stress Concentration and Failure Criteria for Orthotropic and Anisotropic Plates with Circular Openings, Composite Materials: Testing and Design (2nd Conference), ASTM STP 497, p. 363, 1971.

Hagemaier, D. J., Nondestructive Inspection and Quality Control - End-Product Nondestructive Evaluation of Adhesive-Bonded Metal Joints, Section 9, in *Adhesives and Sealants*, Volume 3, *Engineered Materials Handbook*, pp. 729–776, ASM International, Warrendale, PA, 1990.

Halpin, J. C., *Primer on Composite Materials-Analysis*, Technomic, Lancaster, PA, 1984.

Hart-Smith, L. J., Adhesive Bonded Double-Lap Joints, NASA CR-112235, January 1973.

Hart-Smith, L. J., Adhesive Bonded Scarf and Stepped-Lap Joints, NASA CR-112237, January 1973.

Hart-Smith, L. J., Adhesive Bonded Single-Lap Joints, NASA CR-112236, January 1973.

Hart-Smith, L. J., Adhesive Layer Thickness and Porosity Criteria for Bonded Joints, Douglas Aircraft Company, McDonnell Douglas Corporation, US Air Force Technical Report AFWAL-TR-82-4172, December 1982.

Hart-Smith, L. J., Adhesive-Bonded Double-Lap Joints, NASA CR-112235, January 1973.

Hart-Smith, L. J., Analysis and Design of Advanced Composite Bonded Joints, NASA CR-2218, 1974.

Hart-Smith, L. J., Bolted Joints in Graphite-Epoxy Composites, NASA CR-144899, June 1976.

Hart-Smith, L. J., Design and Analysis of Bolted and Riveted Joints in Fibrous Composite Structures, Douglas Paper 7739, Proceedings of the International Symposium on Joining and Repair of Fibre-Reinforced Plastics, at the Imperial College, London, 10–11 September 1986.

Hart-Smith, L. J., Design Methodology for Bonded-Bolted Composite Joints, AFWAL-TR-81-3154, February 1982.

Hart-Smith, L. J., Effect of Flaws and Porosity on Strength of Adhesive Bonded Joints, Douglas Paper 7388, Presented to 29th Annual SAMPE Symposium and Technical Conference, 3–5 April 1984.

Hart-Smith, L. J., Mechanically-Fastened Joints in Graphite-Epoxy Laminates – Phenomenological Considerations and Simple Analyses, Douglas Paper 6748A, presented to Fourth Conference on Fibrous Composites in Structural Design, San Diego, CA, November 1978.

Hart-Smith, L. J., Non-Classical Adhesive-Bonded Joints in Practical Aerospace Construction, NASA CR-112238, January 1973.

Hart-Smith, L. J., Simplified Estimation of Stiffness and Biaxial Strengths for Design of Laminated Carbon-Epoxy Composite Structures, Douglas Aircraft Company, Paper 7548, Presented to Seventh DoD/NASA Conference on Fibrous Composites in Structural Design, Denver, CO, pp. 17–20, June 1985.

Hart-Smith, L. J., The Analysis and Design of Bonded and Bolted Joints for Lightweight Structures, CRC-AS Workshop, Sydney, 17 February 1993.

Hart-Smith, L. J., The Design of Efficient Bolted and Riveted Fibrous Composite Structures, Douglas Paper 8335, presented to TTCP sponsored Workshop organized by U.S. Naval Research Laboratories on Mechanical Joints in Composites, La Jolla, CA, July 1989.

Hart-Smith, L. J. and Heslehurst, R. B., Designing for Repairability, in *ASM Handbook, Volume 21: Composites*. Materials Park, OH, pp. 872–884, December 2001.

Hart-Smith, L. J., Ochsner, R. W. and Radecky, R. L., Surface Preparation of Fibrous Composites for Adhesive Bonding or Painting, *Douglas Services Magazine*, 41, First Quarter 1984.

Hart-Smith, L. J., Wong, S. B. and Brown, D. L., Surface Preparations for Ensuring that the Glue Will Stick in Bonded Composite Structures, McDonnell Douglas Paper MDC 93K0126, presented to 10th DOD/NASA/FAA Conference on Fibrous Composites in Structural Design, Hilton Head Island, SC, November 1–4, 1993, also published in Peters, S. (Ed.), *Handbook of Composites*, 2nd Edition, pp. 667–685, 1998.

Herakovich, C. T., *Mechanics of Fibrous Composites*, Wiley, Chichester, UK, 1998.

Heslehurst, R. B., Analysis and Modelling of Damage and Repair of Composite Materials in Aerospace, in Bull, J. W. (Ed.), Chapter 2, *Numerical Analysis and Modelling of Composite Materials*, Blackie Academic and Professional Publ., Glasgow, pp. 27–59, 1996.

Heslehurst, R. B., Application of a Simple Hole Damage Stress Analysis Method to Composite Airframes: Case Studies, *Journal of Composite Structures*, 27, 253–259, 1994.

Heslehurst, R. B., Challenges in the Repair of Composite Structures – Part I, *The SAMPE Journal*, 33, 5, 11–16, September/October 1997.

Heslehurst, R. B., Challenges in the Repair of Composite Structures – Part II, *The SAMPE Journal*, 33, 6, 16–21, November/December 1997.

Heslehurst, R. B., Composite Structural Repairs an Engineering Approach, 39th International SAMPE Symposium, Covina, CA, May 1994.

Heslehurst, R. B., *Defects and Damage in Composite Materials and Structures*, CRC Press, Boca Raton, FL, 2014.

Heslehurst, R.B., *Design and Analysis of Structural Joints with Composite Materials*, DEStech Publications, Lancaster, PA, 2013.

Heslehurst, R. B., Application and Interpretation of a Portable Holographic Interferometry Testing System in the Detection of Defects in Structural Materials, PhD Thesis, School of Aerospace and Mechanical Engineering, University College, UNSW, 1999.

Heslehurst, R. B., Estimating Adhesive Shear Stress/Strain Properties, Proceedings of the SAMPE 2001 International Symposium and Exhibition, Long Beach, CA, pp. 380–388, May 2001.

Heslehurst, R. B., Evaluation of Damage Analysis Techniques for Composite Aircraft Structures, Masters of Engineering Thesis, Royal Melbourne Institute of Technology, 1991.

Heslehurst, R. B., Repair of Delamination Damage a Simplified Approach, Proceedings of the 41st International SAMPE Symposium, Covina, CA, 24–28 March 1996.

Heslehurst, R. B., *Repair, International Encyclopaedia of Composites*, 2nd Edition, SAMPE, Covina, CA, 2000.

Heslehurst, R. B., Scarf Repair – Is it Best to have the Big Ply Down First or the Little Ply Down First? Technical Note, *Composites Institute of Australia – Quarterly Newsletter*, March 2009.

Heslehurst, R. B., Structural Repair Methodology for Wind Turbine Blades, Composites Australia – Annual Conference, Sanctuary Cove, Queensland, 22–23 April 2015.

Heslehurst, R. B., The Accuracy of Simple Damage Analysis Methods in Composite Structures, Proceedings of the 37th International SAMPE Symposium, Anaheim, CA, pp. 321–332, March 1992.

Heslehurst, R. B., Which is Best, a Bolted or a Bonded Patch? Presented at the SAE TOPTEC Series Conference, Composite Structural Repairs, Seattle, WA, November 1993.

Heslehurst, R. B., What is the Best Way to Repair Delaminations? Technical Note, *Abaris Training Resources Newsletter*, January 1999.

Heslehurst, R. B. and Forte, M., Repair Engineering and Design Considerations, in *ASM Handbook, Volume 21: Composites*. Materials Park, OH, pp. 885–892, December 2001.

Heslehurst, R. B. and Scott, M. L., Review of Defects and Damage Pertaining to Aircraft Composite Structures, *Journal of Composite Polymers*, 3, 2, pp. 103–133, 1990.

Heslehurst, R. B., Baird, J. P., Williamson, H. M. and Clark, R. K., Can Aging Adhesively Bonded Joints Be Found? Proceedings of the 41st SAMPE International Symposium and Exhibition, Anaheim, CA, pp. 925–935, March 1996.

Heslehurst, R. B., Vaughan, M. J., Baird, J. P. and Clark, R. K., Experimental Validation of Delamination Buckling Characteristics, Proceedings of the 6th Australian Aeronautical Conference, Melbourne, pp. 199–204, March 1995.

Heslehurst, R. B., Vaughan, M. J., Baird, J. P., Williamson, H. M. and Clark, R. K., Delamination Buckling Characteristics, Presented at the 5th Australian Aeronautical Conference, Melbourne, Australia, September 1993.

Heslehurst, R. B., Dorworth, L. C. and Hoke, M. J., Comparison of Two Scarf Repair Configurations, *International SAMPE Symposium and Exhibition (Proceedings)*, 45, I, pp. 57–63, 2000.

Hinton, M. J., Kaddour, A. S. and Soden, P. D., *Failure Criteria in Fibre-Reinforced-Polymer Composites: The World-Wide Failure Exercise*, Elsevier Science, Sydney, Australia, 2004.

Hoffman, G. A. and Konishi, D. Y., Characterizations of Manufacturing Flaws in Graphite/Epoxy, Los Angeles Aircraft Division of Rockwell Int, AMMRC-MS-77-5, 1977.

Hoskin, B. C. and Baker, A. A. (Eds.), *Composite Materials for Aircraft Structures*, AIAA Education Series, AIAA, New York, NY, 1986.

Hsu, D. K., Non-Destructive Inspection of Composite Structures: Methods and Practice, 17th World Conference on Non-Destructive Testing, Shanghai, China, 25–28 October 2008.

Hull, D., *An Introduction to Composite Materials*, Cambridge University Press, Cambridge, UK, 1981.

Johnson, W. and Ghosh, S. K., Some Physical Defects Arising in Composite Material Fabrication – Review, *Journal of Materials Science*, 16, pp. 285–301, 1981.

Jones, R. M., *Mechanics of Composite Materials*, 2nd Edition, CRC Press, Boca Raton, FL, 1998.

Kai, S. and Heslehurst, R. B., Scarf Repair Lay-Up Orientation Study, *Journal of Advanced Materials*, 40, 2, pp. 65–71, 2008.

Karbhari, V., *Non-Destructive Evaluation (NDE) of Polymer Matrix Composites*, Woodhead Publishing, Cambridge, UK, 2013.

Kim, H. and Kedward, K. T., *Joining and Repair of Composite Structures* (STP 1455), ASTM, West Conshohocken, PA, 2004.

Kirschke, L., Damage Mechanisms in CFRP Laminates with Defects, Contributions on the Properties of Carbon Fibre Reinforced Composites, ESA-TT-849, pp. 155–204, July 1984.

Kollar, L. and Springer, G., *Mechanics of Composite Structures*, Cambridge University Press, Cambridge, UK, 2009.

Konishi, D. Y. and Lo, K. H., Flaw Criticality of Graphite/Epoxy Structures. Non-Destructive Evaluation and Flaw Criticality for Composite Materials, ASTM STP 696, pp. 125–144, October 1978.

Labor, J. D., Repair Procedures for Composite Sinewave Substructure, 16th National SAMPE Technical Conference, p. 129, 9–11 October 1984.

Landrock, A. H., *Adhesives Technology Handbook*, Noyes Publications, New Jersey, 1985.

Lekhnitski, S. G., *Anisotropic Plates*, Gordon and Breach, New York, NY, 1968.

Lubin, G. *Handbook of Composites*, 1st Edition, Van Norstrand Reinhold, New York, NY, 1982.

Luhmann, T., *Close Range Photogrammetry: Principles, Techniques and Applications*, Whittles Publishing, Dunbeath, Scotland, 2006.

Mallick, P. K., *Composites Engineering Handbook*, M. Dekker, New York, 1997.

Masters, J. E., Characterization of Impact Damage Development in Graphite/Epoxy Laminates, Fractography of Modern Engineering Materials, Composites and Metals, ASTM STP 948, pp. 238–258, 1987.

Matthews, F. L., *Joining Fibre-Reinforced Plastics*, Elsevier Applied Science Publishers, London, UK, 1986.

Matthews, F. L., Wong, C. M. and Chryssafitis, S., Stress Distribution Around a Single Bolt in Fibre-Reinforced Plastic, *Composites*, 13, 3, 316, 1982.

Miracle, D. B. and Donaldson, S. L., *ASM Handbook, Volume 21: Composites*, ASM International, Ohio, 2018.

MMPDS, Metallic Materials Properties Development and Standardization, SAE International, Warrendale, PA, 2019.

Mouritz, A. P., Repair of Composite Ship Structures, Aeronautical and Maritime Research Laboratory, DSTO, Chapter for Section 9, Product Reliability, Maintainability and Repair, in *ASM International's Engineered Materials Handbook, Volume 1: Composites*, 2nd Edition, 2000.

Mouritz, A. P., Repair of Composite Ship Structures, Aeronautical and Maritime Research Laboratory, DSTO, in ASM Handbook, Volume 21: Composites. Materials Park, OH, pp. 889–907, December 2001.

Myhre, S. H. and Beck, C. E., Repair Concepts for Advanced Composite Structures, *Journal of Aircraft*, 16, 10, pp. 720–728, 1979.

Myhre, S. H. and Labor, J. D., Repair of Advanced Composite Structures, *Journal of Aircraft*, 18, 7, pp. 546–552, 1981.

Niu, M. C. Y., *Airframe Stress Analysis and Sizing*, 2nd Edition, Conmilit Press, Hong Kong, 1999.

Niu, M. C. Y., *Airframe Structural Design: Practical Design Information and Data on Aircraft Structures*, Conmilt Press, Hong Kong, 1988.

Niu, M. C. Y., *Composite Airframe Structures*, Conmilit Press, Hong Kong, 1992.

No Author, *Advanced Composite Repair Guide*, NOR 82-60, Northrop Corporation, Hawthorne, March 1982.

No Author, *Aircraft Bonded Structure*, IAP Inc., Training Manual, Casper, 1985.

No Author, *Composite Materials Handbook 17*, Volumes 1, 2, 3 and 6, SAE International, Warrendale, PA, 2017.

Nuismer, R. J. and Whitney, J. M., Uniaxial Failure of Composite Laminates Containing Stress Concentrations, in Sendeckyj, G. P. (Ed.), *Fracture Mechanics of Composite*, ASTM STP 593, pp. 117–142, Philadelphia, PA, 1975.

O'Brien, T. K., Characterization of Delamination Onset and Growth in a Composite Laminate, Damage in Composite Materials: Basic Mechanisms, Accumulation, Tolerance and Characterization, ASTM STP 775, pp. 140–167, 1982.

Odi, R. A. and Friend, C. M., An Improved 2D Model for Bonded Composite Joints, *International Journal of Adhesion and Adhesives*, 24, 5, pp. 389–405, 2004.

Pagano, N. J. and Pipes, R. B., *Interlaminar Response of Composite Materials*, Elsevier Science Publishers, Amsterdam, the Netherlands, 1989.

Perl, D., Depot Repairs of F/A-18 Composite Aircraft Structures, Naval Aviation Depot Report, North Island, July 1983.

Peters, P., Maintenance of Fibre Reinforced Plastics on Aircraft, Department of Aviation, Aircraft Maintenance – Text 5, AGPS, Canberra, 1986.

Peters, W. H. and Ranson, W. F., Digital Imaging Techniques in Experimental Stress Analysis, *Optical Engineering* 21, p. 3, 1982.

Potter, J. M., *Fatigue in Mechanically Fastened Composite and Metallic Joints*, ASTM Special Technical Publication STP-927, American Society for Testing and Materials, Philadelphia, PA, 1986.

Puralow, D., Fractographic Analysis of Failures in CFRP, *Characterization, Analysis and Significance of Defects in Composite Materials*, AGARD-CP-355, April 1983.

Ramamurthy, G., Mechanics of Debond Growth in Adhesively Bonded Composite Joints, Master of Engineering Mechanics, University of Missouri-Rolla, 1986.

Ratwani, M. M., Repair Options for Airframes, Section 4, Aging Aircraft Fleets: Structural and Other Subsystem Aspects, RTO Lecture Series 218, 13–16 November 2000.

Ratwani, M. M., Repair Types, Procedures – Part I, RTO-EN-AVT-156, 2010.

Reifsnider, K. L. (Ed.), *Damage in Composite Materials: Basic Mechanisms, Accumulation, Tolerance, and Characterization*, ASTM STP 775, 1982.

Reifsnider, K.L., Non-Destructive Test Methods for Composite Structures, 30th National SAMPE Symposium, 19–21 March 1985.

Reynolds, W. N., Nondestructive Testing (NDT) of Fibre-Reinforced Composite Materials, *Materials and Design*, 5, 6, December/January 1985.

Roh, H. S., *Repairs of Composite Structures*, ProQuest, UMI Dissertation Publishing, Ann Arbor, MI, 2012.

Rubin, A. M. and Perl, D. R., Composite Structures Repair Development for the F/A-18E/F Aircraft, Proceedings of the 40th International SAMPE Symposium, pp. 154–167, 8–11 May 1995.

Schmidt, T., Tyson, J. and Galanulis, K., Full-Field Dynamic Displacement and Strain Measurement Using Advanced 3D Image Correlation Photogrammetry: Part I, *Experimental Techniques*, 27, 3, pp. 47–50, 2003.

Schwartz, M. M. *Joining of Composite-Matrix Materials*, ASM International, Warrendale, PA, 1994.

Shackleford, J. F., *Introduction to Materials Science for Engineers*, 3rd Edition, Macmillan, Sydney, Australia, 1992.

Soutis, C. and Hu, F. Z. A 3-D Failure Analysis of Scarf Patch Repaired CFRP Plates, AIAA/ASME/ASCE/AHS/ASC Structures, Structural Dynamics, and Materials Conference and Exhibit, 39th, and AIAA/ASME/AHS Adaptive Structures Forum, Long Beach, CA, 20–23 April 1998.

Stone, D. E. W. and Clarke, B., In-Service NDI of composite Structures: An Assessment of Current Requirements and Capabilities, in Chapter 18, *Characterization, Analysis and Significance of Defects in Composite Materials*, AGARD-CP-335, April 1983.

Street, K. N., Defect Criticality in Composite Structures: Delaminations, TTCP Subgroup P – Materials TTP 4, Interim Report, April 1987.

Strong, A. B., *Fundamentals of Composites Manufacturing: Materials, Methods and Applications*, 2nd Edition, Society of Manufacturing Engineers, Dearborn, MI, 2007.

Summerscales, J. (Ed.), *Non-Destructive Testing of Fibre-Reinforced Plastics Composites*, Volume 1, Elsevier Applied Science, London, UK, 1987.

Sutter, D. A., Three-Dimensional Analysis of a Composite Repair and the Effect of Overply Shape Variation on Structural Efficiency, Master of Aeronautical Engineering Thesis, Air Force Institute of Technology, Air University, March 2007.

Tan, S. C., *Stress Concentrations in Laminated Composites*, CRC Press, Boca Raton, FL, 1994.

The Institution of Mechanical Engineers, International Conference on Joining and Repair of Plastics and Composites, London, UK, pp. 16–17, March 1999.

Tsai, S. W., *Composite Design*, 4th Edition, Thick Composites, Dayton, OH, 1988.

Tsai, S. W. and Hahn, H. T., *Introduction to Composite Materials*, Technomic, Lancaster, PA, 1980.

Weeton, J. W., Peters, D. M. and Thomas, K. L. (Eds.), *Engineer's Guide to Composite Materials*, ASM International, Dayton, OH, 1987.

Wegman, R. F., *Surface Preparation Techniques for Adhesive Bonding*, Noyes Publications, Saddle River, NJ, 1989.

Whitney, J. M., *Structural Analysis of Laminated Anisotropic Plates*, Technomic, Lancaster, PA, 1987.

Whitney, J. M. and Nuismer, R. J., Stress Fracture Criteria for Laminated Composites Containing Stress Concentrations, *Journal of Composite Materials*, 8, pp. 253–265, 1974.

Whitney, J. M., Daniel, I. M. and Pipes, R. B., Experimental Mechanics of Fibre Reinforced Composite Materials, *SESA Monograph*, #4, 1982.

Wicker, H., Composite Structures Repair, ICAS-82-3.8.2, Proceedings of the 13th Congress of the International Council of the Aeronautical Sciences, Seattle, WA, p. 1386, 22–27 August 1982.

Wilkins, D. J., Engineering Significance of Defects in Composite Structures, in Chapter 20, *Characterization, Analysis and Significance of Defects in Composite Materials*, AGARD-CP-335, 1983.

OEM DOCUMENTS

Falcon 900, Inspection of External Composite Material Components, August 1988.

Falcon 900, Repair Manual Standard Practices Concerning Composite Material Components, August 1987.

General Dynamics, F-111 Repairs for Reinforced Plastic Assemblies, T.O. F- 111C, Structural Repair Manual.

McDonnell Douglas Hornet, Maintenance and Repair Training Notes, Section 9, Composite Repair.

Pilatus, PC9, Structural Repair Manual, Repairs to Glass-Fibre Laminates, Section 51-70-02, April 1987.

Sikorsky Aircraft, Blackhawk, Fibreglass Reinforced Plastics Repair-Training Course Notes.

United Technologies Sikorsky Aircraft, Composite Repair, Training Manual, REV. 4, 6-84-TRW.

NASA PUBLICATIONS

NASA-CR-134525, Mazzio, V. F,, Mehan, R. L. and Mullin, J. V., Basic Failure Mechanisms In Advanced Composites, 1973.

NASA-CR-159056, Stone, R., Repair Techniques for Graphite-Epoxy Structures for Commercial Transport Applications, January 1983.

NASA-CR-169708, Pipes, R. B. and Adkins, D. W. Strength and Mechanics of Bonded Scarf Joints for Repair of Composite Materials [Final Report, June 1981–May 1982] (24 Composite Materials (AH) No.; NAS 1.26:169708; CCM-82-14; Pagination 190P).

NASA-CR-3794, Jones, J. S. and Graves, S. R., Repair Techniques for Celion/LARC-160 Graphite/Polyimide Composite Structures, June 1984.

NASA-CR-4733, Flynn, B. W., Bodine, J. B., Dopker, B., Finn, S. R., Griess, K. H., Hanson, C. T., Harris, C. G., Nelson, K. M., Walker, T. H., Kennedy, T. C. and Nahan, M. F., Advanced Technology Composite Fuselage – Repair and Damage Assessment Supporting Maintenance, April 1997.

NASA-TM-100281, Chamis, C. C., Simplified Procedures for Designing Composite Bolted Joints, 1988.

NASA-TM-76947, Schnetze, R. and Hillger, W., Recognizing Defects in Carbon-Fibre Reinforced Plastics, September 1982.

NASA-TM-84505, Deaton, J. W., A Repair Technology Program at NASA on Composite Materials, August 1982.

NASA-TM-86038, Ko, W. L., Stress Concentrations Around a Small Circular Hole in the HiMAT Composite Plate, December 1985.

MILITARY DOCUMENTS

AFFDL-TM-74-216-FBT, Noble, R. A., Acoustic Emission for Damage Detection in USAF Composite Structural Components, January 1975.

AFFDL-TR-78-79, McCarty, J. E., Horton, R. E., Locke, M. C., Mayberry, R. Z. and Satterthwait, M. L., Repair of Bonded Primary Structure, June 1978.

AFML-TR-76-201, McCarty, J. E., Horton, R. E., Locke, M. C., Satterthwait, M. and Parashar, B. D., Adhesive Bonded Aerospace Structures Standardized Repair Handbook, December 1976.

AFML-TR-76-65, McCarty, J. E., Horton, R. E., Locke, M. C., Satterthwait, M. and Parashar, B. D., Adhesive Bonded Aerospace Structures Standardized Repair Handbook, May 1976.

AFML-TR-76-81-Part I-IV, Reifsnider, K. L., Henneke II, E. G. and Stinchcomb, W. W., Defect-Property Relationships In Composite Materials, June 1978.

AFML-TR-77-206, Horton, R. E. and McCarty, J. E., Adhesive Bonded Aerospace Structures Standardized Repair Handbook, December 1977.

AFML-TR-79-4087, Chang, F. H., Bell, J. R., Gardner, A. H., Handley, G. P. and Fisher, C. P., In Service Inspection of Advanced Composite Aircraft Structure, February 1979.

AFRL-RX-WP-JA-2016-0028, Lindgren, E. A., Brausch, J. and Buynak, C., Recent and Future Enhancements in NDI for Aircraft Structures, November 2015.

AFRL-RX-WP-TR-2008-4373, Brausch, J., Butkus, L., Campbell, D., Mullis, T., and Paulk, M., Recommended Processes and Best Practices for Nondestructive Inspection (NDI) of Safety-of-Flight Structures, October 2008.

AFWAL-TR-81-3041-Vol 3, Garbo, S. P. and Ogonowski, J. M., Effect of Variances and Manufacturing Tolerances on the Design Strength and Life of Mechanically Fastened Composite Joints, *Bolted Joint Stress Field Model (BJSFM) Computer Program User's Manual*, Volume 3, April 1981.

AFWAL-TR-84-4029, Chang, F.-K., Scott, R. A., Springer, G. S., Strength of Bolted Joints in Laminated Composites, Final report for period, June 1983–December 1983.

AFWAL-TR-87-3104, Hinkle, T. and Van ES, J., Battle Damage Repair of Composite Structures, March 1988.

AFWAL-TR-87-4109, Martin, W., Advanced Field Level Repair Materials Technology for Composite Structures, October 1987.

ARL-RR-7, Chalkley, P. D., Mathematical Modelling of Bonded Fibre-Composite Repairs to Aircraft, May 1993.

ARL-STRUC-REPORT-386, Jones, R., Callinan, R. J. and Aggarwal, K. C., Stress Analysis of Adhesively Bonded Repairs to Fibre Composite Structures, March 1981.

ARL-TM-379, Scott, I. G. and Scala, C. M., NDI of Composite Materials, Aeronautical Research Labs Melbourne (Australia), July 1981.

DSTO-RR-0317, Wang, C. H. and Gunnion, A., Design Methodology for Scarf Repairs to Composite Structures, Defence Science and Technology Organisation, DSTO, August 2006.

DSTO-TN-0275, Shah Khan, M. Z. and Grabovac, I., Repair of Damage to Marine Sandwich Structures: Part II – Fatigue Testing, DSTO Aeronautical and Maritime Research Laboratory, Melbourne Victoria, May 2000.

DSTO-TR-0608, Chalkley, P. and van den Berg, J., On Obtaining Design Allowables for Adhesives Used in the Bonded Composite Repair of Aircraft, DSTO Aeronautical and Maritime Research Laboratory, Melbourne Victoria, January 1998.

DSTO-TR-0736, Thomson, R., Luescher, R. and Grabovac, I., Repair of Damage to Marine Sandwich Structures: Part I – Static Testing Aeronautical and Maritime Research Laboratory, Melbourne Victoria, October 1996.

MIL-HDBK-17, see Composite Materials Handbook 17.

MIL-HDBK-337, Department of Defense Handbook, Adhesive Bonded Aerospace Structure Repair, 01 December 1982.

MIL-HDBK-6870A, Department of Defense Handbook, Handbook Inspection Program Requirements Nondestructive for Aircraft and Missile Materials and Parts, 14 February 2012.

MIL-HDBK-691B, Department of Defense Handbook, Adhesive Bonding, 12 March 1987.

MIL-HDBK-754(AR), Military Handbook, Plastic Matrix Composites with Continuous Fiber Reinforcement, 19 September 1991.

MIL-HDBK-787, Military Standardization Handbook Non-Destructive Testing Methods of Composite Materials-Ultrasonics, 1 April 1988.

MIL-HDBK-793(AR), Department of Defense Handbook, Nondestructive Active Testing Techniques for Structural Composites, 6 November 1989.

MIL-HDBK-83377, Department of Defense Handbook, Adhesive Bonding (Structural) For Aerospace and other Systems, Requirements For, 31 December Y.

NADC-77109-30, Watson, J. B., Glaeser, D. A., Harvey, F. L., Fukimoto, W. T. and Padilla, V. E., Bolted Field Repair of Composite Structures, March 1979.

NADC-80032-60, McGovern, S. A., NDI Survey of Composite Structures, Naval Air Systems Command, Maintenance Technology Program, January 1980.

NADC-80126-60 Weiss, J., Composite Repair System with Long-Term Latency, Naval Air Development Center, Warminster, PA, November 1981.

NADC-81063-60, Bohlmann, R. E., Renier, G. D. and Horton, D. K., Bolted Repair Analysis Methodology, McDonnell Aircraft Corp., December 1982.

NAEC-MISC-92-0366, Nesterok, D. W., Advanced Composite In-Service Damage Assessment/NDI Development Program, September 1978.

RAE, Potter, R. T. and Dorey, G., The Effects and Structural Significance of Defects and Damage in Composite Materials, Farnborough, 1984.

RAE-TM-Mat/Str-1014, Dorey, G., Mousley, R. F. and Potter, R. T., Significance of Defects in Composite Structures (The): A Review of Current Research and Requirements, April 1983.

RAE TM MAT-394, Bishop, S. M., Significance of Defects on the Failure of Fibre Composites (The): (A Review of Research in the UK), March 1982.

WL-TR-92-4069, Kuhbander, R. J., Characterization of EA9394 Adhesive for Repair Applications, July 1994.

WL-TR-92-4084, Tan, S. C., Analysis of Bolted and Bonded Composite Joints, September 1992.

WL-TR-97-3105, Boyd, K. L., Krishnan, S., Litvinov, A., Elsner, J. H., Ratwani, M. M., Harter, J. A. and Glinka, G., Development of Structural Integrity Analysis Technologies for Aging Aircraft Structures: Bonded Composite Patch Repair and Weight Function Methods, July 1997.

WRDC-TR-90-4035, Askins, D. R., Kuhbander, R., Saliba, S., Griffen, C., Lawless, G. W., McKiernan, J. and Behme, A., Composites Supportability Rapid Test and Evaluation, May 1990.

FEDERAL AVIATION ADMINISTRATION REPORTS

AC 145-6, Repair Stations for Composite and Bonded Aircraft Structure U.S. Department of Transportation, Federal Aviation Administration, 15 November 1996 – Cancelled.

AC 20-107B, Composite Aircraft Structure, U.S. Department of Transportation, Federal Aviation Administration, 8 September 2009.

DOT/FAA/AR-00/44, Raju, K. S., Liew, J. and Smith, B. L., Impact Damage Characterization and Damage Tolerance of Composite Sandwich Airframe Structures-9259, Tomblin J.S., U.S. Department of Transportation, Federal Aviation Administration, Office of Aviation Research Washington DC, January 2001.

DOT/FAA/AR-00/46, Ahn, S.-H. and Springer, G. S., Repair of Composite Laminates, U.S. Department of Transportation, Federal Aviation Administration, Office of Aviation Research Washington DC, December 2000.

DOT/FAA/AR-01/08, Bardis, J. and Kedward, K., Effects of Surface Preparation on Long-Term Durability of Composite Adhesive Bonds, U.S. Department of Transportation, Federal Aviation Administration, Office of Aviation Research Washington DC, April 2001.

DOT/FAA/AR-01/56, Davidson, B. D., A Predictive Methodology for Delamination Growth in Laminated Composites Part II-9269, U.S. Department of Transportation, Federal Aviation Administration, Office of Aviation Research Washington DC, October 2001.

DOT/FAA/AR-02/121, Shyprykevich, P., Tomblin, J., Ilcewicz, L., Vizzini, A. J., Lacy, T. E. and Hwang, Y., Guidelines for Analysis, Testing, and Non-destructive Inspection of Impact- Damaged Composite Sandwich Structures, U.S. Department of Transportation, Federal Aviation Administration, Office of Aviation Research Washington DC, March 2003.

DOT/FAA/AR-03/74, Tomblin, J. S., Salah, L., Welch, J. M. and Borgman, M. D., Bonded Repair of Aircraft Composite Sandwich Structures, U.S. Department of Transportation, Federal Aviation Administration, Office of Aviation Research Washington DC, February 2004.

DOT/FAA/AR-05/12, Zhu, Y. and Kedward, K., Methods of Analysis and Failure Predictions for Adhesively Bonded Joints of Uniform and Variable Bondline Thickness, U.S. Department of Transportation, Federal Aviation Administration, Office of Aviation Research Washington DC, May 2005.

DOT/FAA/AR-97/87, Davidson, B. D., A Predictive Methodology for Delamination Growth in Laminated Composites Part I-Theoretical Development and Preliminary Experimental Results, U.S. Department of Transportation, Federal Aviation Administration, Office of Aviation Research Washington DC, April 1998.

DOT/FAA/CT-93/79, Rice, R., Francini, R., Rabman, S., Rosenfeld, M., Rust, B., Smith, S. and Brook, D., Effects of Repair on Structural Integrity, U.S. Department of Transportation, Federal Aviation Administration Technical Center, Atlantic City International Airport, NJ, December 1993.

Index

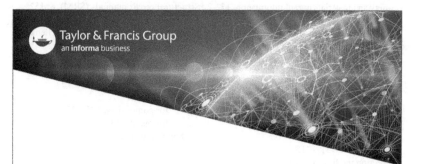

Milton Keynes UK
Ingram Content Group UK Ltd.
UKHW040059071024
449327UK00019B/657

9 780367 779962